WHAT EVERY ENGINEER
SHOULD KNOW ABOUT

MODELING AND SIMULATION

WHAT EVERY ENGINEER SHOULD KNOW
A Series

Series Editor*

Phillip A. Laplante

Pennsylvania State University

1. What Every Engineer Should Know About Patents, *William G. Konold, Bruce Tittel, Donald F. Frei, and David S. Stallard*
2. What Every Engineer Should Know About Product Liability, *James F. Thorpe and William H. Middendorf*
3. What Every Engineer Should Know About Microcomputers: Hardware/Software Design, A Step-by-Step Example, *William S. Bennett and Carl F. Evert, Jr.*
4. What Every Engineer Should Know About Economic Decision Analysis, *Dean S. Shupe*
5. What Every Engineer Should Know About Human Resources Management, *Desmond D. Martin and Richard L. Shell*
6. What Every Engineer Should Know About Manufacturing Cost Estimating, *Eric M. Malstrom*
7. What Every Engineer Should Know About Inventing, *William H. Middendorf*
8. What Every Engineer Should Know About Technology Transfer and Innovation, *Louis N. Mogavero and Robert S. Shane*
9. What Every Engineer Should Know About Project Management, *Arnold M. Ruskin and W. Eugene Estes*
10. What Every Engineer Should Know About Computer-Aided Design and Computer-Aided Manufacturing: The CAD/CAM Revolution, *John K. Krouse*
11. What Every Engineer Should Know About Robots, *Maurice I. Zeldman*
12. What Every Engineer Should Know About Microcomputer Systems Design and Debugging, *Bill Wray and Bill Crawford*
13. What Every Engineer Should Know About Engineering Information Resources, *Margaret T. Schenk and James K. Webster*
14. What Every Engineer Should Know About Microcomputer Program Design, *Keith R. Wehmeyer*
15. What Every Engineer Should Know About Computer Modeling and Simulation, *Don M. Ingels*
16. What Every Engineer Should Know About Engineering Workstations, *Justin E. Harlow III*
17. What Every Engineer Should Know About Practical CAD/CAM Applications, *John Stark*
18. What Every Engineer Should Know About Threaded Fasteners: Materials and Design, *Alexander Blake*
19. What Every Engineer Should Know About Data Communications, *Carl Stephen Clifton*
20. What Every Engineer Should Know About Material and Component Failure, Failure Analysis, and Litigation, *Lawrence E. Murr*
21. What Every Engineer Should Know About Corrosion, *Philip Schweitzer*
22. What Every Engineer Should Know About Lasers, *D. C. Winburn*
23. What Every Engineer Should Know About Finite Element Analysis, *John R. Brauer*

*Founding Series Editor: William H. Middendorf

24. What Every Engineer Should Know About Patents: Second Edition, *William G. Konold, Bruce Tittel, Donald F. Frei, and David S. Stallard*
25. What Every Engineer Should Know About Electronic Communications Systems, *L. R. McKay*
26. What Every Engineer Should Know About Quality Control, *Thomas Pyzdek*
27. What Every Engineer Should Know About Microcomputers: Hardware/Software Design, A Step-by-Step Example, Second Edition, Revised and Expanded, *William S. Bennett, Carl F. Evert, and Leslie C. Lander*
28. What Every Engineer Should Know About Ceramics, *Solomon Musikant*
29. What Every Engineer Should Know About Developing Plastics Products, *Bruce C. Wendle*
30. What Every Engineer Should Know About Reliability and Risk Analysis, M. Modarres
31. What Every Engineer Should Know About Finite Element Analysis: Second Edition, Revised and Expanded, *John R. Brauer*
32. What Every Engineer Should Know About Accounting and Finance, *Jae K. Shim and Norman Henteleff*
33. What Every Engineer Should Know About Project Management: Second Edition, Revised and Expanded, *Arnold M. Ruskin and W. Eugene Estes*
34. What Every Engineer Should Know About Concurrent Engineering, *Thomas A. Salomone*
35. What Every Engineer Should Know About Ethics, *Kenneth K. Humphreys*
36. What Every Engineer Should Know About Risk Engineering and Management, *John X. Wang and Marvin L. Roush*
37. What Every Engineer Should Know About Decision Making Under Uncertainty, *John X. Wang*
38. What Every Engineer Should Know About Computational Techniques of Finite Element Analysis, *Louis Komzsik*
39. What Every Engineer Should Know About Excel, *Jack P. Holman*
40. What Every Engineer Should Know About Software Engineering, *Phillip A. Laplante*
41. What Every Engineer Should Know About Developing Real-Time Embedded Products, *Kim R. Fowler*
42. What Every Engineer Should Know About Business Communication, *John X. Wang*
43. What Every Engineer Should Know About Career Management, *Mike Ficco*
44. What Every Engineer Should Know About Starting a High-Tech Business Venture, *Eric Koester*
45. What Every Engineer Should Know About MATLAB® and Simulink®, *Adrian B. Biran with contributions by Moshe Breiner*
46. Green Entrepreneur Handbook: The Guide to Building and Growing a Green and Clean Business, *Eric Koester*
47. Technical Writing: A Practical Guide for Engineers and Scientists, *Phillip A. Laplante*
48. What Every Engineer Should Know About Cyber Security and Digital Forensics, *Joanna F. DeFranco*
49. What Every Engineer Should Know About Modeling and Simulation, *Raymond J. Madachy and Daniel X. Houston*
50. What Every Engineer Should Know About Excel, Second Edition, *J.P. Holman and Blake K. Holman*

WHAT EVERY ENGINEER SHOULD KNOW ABOUT

MODELING AND SIMULATION

Raymond J. Madachy
Daniel X. Houston

CRC Press is an imprint of the
Taylor & Francis Group, an **informa** business

CRC Press
Taylor & Francis Group
6000 Broken Sound Parkway NW, Suite 300
Boca Raton, FL 33487-2742

Printed on acid-free paper

International Standard Book Number-13: 978-1-4987-5309-8 (Paperback)
International Standard Book Number-13: 978-1-138-29750-0 (Hardback)

Visit the Taylor & Francis Web site at
http://www.taylorandfrancis.com

and the CRC Press Web site at
http://www.crcpress.com

Dedication

To our families who allowed us the time to write this book.

Contents

Preface

This book presents fundamental concepts and issues in computer modeling and simulation (M&S) in a simple and practical way for engineers, scientists, and managers who wish to apply simulation successfully to their real-world problems. Enabled by computing advances, the field of M&S has become an important industrial tool for decision support, training, forecasting and planning, and marketing. For example, we want to be better informed about the challenging decisions we make about large or complex systems and processes, whether they are physical or organizational. We prefer to have evidence that supports our decisions rather than learning later that we unwittingly chose a lesser option or made a serious mistake due to lack of information. Learning how a new system will respond before building it, or what effects will be seen with changes to an existing dynamic system, can be far more cost effective than performing trials on a real system.

The applications of M&S are growing and will continue to grow. Even those who do not practice M&S are more frequently called upon to engage with modelers and simulation models. Thus, engineers and technical managers increasingly need to be conversant and knowledgeable in the concepts and techniques of M&S.

The merits of the book lie in its broadness, simplicity, and unique approach to the coverage of generic (tool-independent) M&S concepts for engineers. The book provides lessons learned from many years of working on M&S projects. This approach enables engineering practitioners to easily learn, evaluate, and apply various available simulation concepts.

Worked-out examples are included to illustrate the concepts. An example modeling application is continued throughout the chapters to demonstrate the techniques. Readers can also access the example simulation programs. Guidelines and checklists to support the modeling process are provided, and an entire chapter is devoted to an actual modeling case study that was performed in industry.

Audience

In contrast with other M&S books that are directed solely toward practitioners, this book is for other audiences including students of M&S, beginning M&S practitioners, and other stakeholders in M&S projects whether as managers, sponsors, or domain experts. Other textbooks focus primarily on the technical aspects and do not describe the broader trade-offs that a modeler faces. This book explains the larger questions and issues to be addressed.

For students of M&S, operations research, and management science, this book takes the modeler's perspective and offers insights on the considerations that arise in an M&S project. Modeling projects are likely to fail without this broader holistic view of the M&S process.

This book also addresses beginning M&S practitioners, especially those in industry without formal training who may have been asked or seen an opportunity to produce a dynamic model. Simulation software tools have advanced dramatically in recent decades with graphical interfaces, specialized libraries, and open-source software that allow easy entry into the practice of M&S. However, those who attempt to use these software packages without adequate preparation may find that that their models provide little, if any, benefit. They may lack a well-rounded understanding of M&S as a statistical tool for policy analysis and decision support. Six Sigma and other quality practitioners are in this category, even though M&S is not among the basic set of tools it is a powerful one to address the same quality problems. For beginning M&S practitioners, the book discusses the role of a practitioner, knowledge and skills required, the role of M&S in decision support, and the stages of an M&S project engagement.

Another audience for this book includes a wide cross section of engineering, engineering management, and technical professions who are stakeholders in M&S projects. Some may already appreciate the value of good modeling but may want to obtain a better understanding of issues faced in modeling so as to improve their oversight or support for a modeling project. Others may be reluctant to participate in M&S projects due to skepticism about the benefits of modeling given its cost. For this audience, this book provides an introduction to the concepts, methods, and techniques of M&S.

The reader does not need prior M&S experience to understand and utilize this book. However, introductory probability and statistics is necessary for fully understanding M&S and is highly recommended.

Book Contents

The introduction in Chapter 1 explains the basics of M&S. Important terms and key concepts are overviewed. The breadth of engineering applications is discussed. An introductory example showing how basic measures are computed in a simulation helps set the context for subsequent chapters.

In Chapter 2, the stages and activities of M&S are discussed with extensive guidance for carrying out an M&S project successfully. All stages involved in the simulation process are described including stakeholder engagement, system representation, data collection, model construction, output analysis, verification, validation, presentation of results, and making recommendations. It includes tips for working with stakeholders, modeling heuristics, and rules-of-thumb to deal with real-world scenarios. It provides detailed checklists and guidelines to support both new and experienced modelers. The reader will be better prepared to successfully manage, conduct, and use the results of modeling projects to improve systems and processes.

Chapter 3 explains different types of simulation models. The underpinnings of continuous, discrete event and agent-based models are described. It demonstrates the concepts with simple, illustrative models for all the simulation model types. The example models are also made available to readers.

The important statistical nature of M&S is treated with separate chapters on randomness, input analysis, and output analysis. Chapter 4 describes how to represent uncertainty in models. It explains the different types of probability distributions, how to generate random values, and the Monte Carlo technique that is extensively used in simulation.

Chapters 5 and 6 deal with simulation input analysis and output analysis respectively. The importance of randomness is made clear in both contexts. The best practices for data collection and elicitation are described, and how to drive models with stochastic input. The issues and methods for statistically analyzing model outputs are explained. The techniques for hypothesis testing and confidence intervals are also covered with examples.

Chapter 7 details a real-world case study that encapsulates all the major concepts of previous chapters. The M&S project successfully forecasts availability of mission-critical software, and also provides additional benefits to the organization. The case study amply demonstrates the advantages of performing M&S in industry.

Finally, appendices are devoted to simulation tool descriptions and simulation program examples are provided. All source code examples are also available on the Internet.

ACKNOWLEDGMENTS

We would like to extend sincere appreciation to others who helped our book development. Our influential professors in modeling and simulation include Dr. Alan Schneider and Dr. James Bush in system dynamics at University of California, San Diego; Dr. Behrokh Khoshnevis in discrete systems and Dr. Alex Loewenthal in statistics of simulation at University of Southern California; and Dr. Gerald Mackulak of Arizona State University.

Special thanks to Dr. Behrokh Khoshnevis for the vision of the EZSIM framework and ongoing collaboration with Dr. Kurt Palmer at USC on *Discrete Systems Simulation* from which our book drew heavily.

We have also learned from our numerous colleagues over the years in simulation research and practical applications. In particular, we have been fortunate to have been involved with an active systems and software process modeling community.

Many thanks to our technical peer reviewers for their detailed constructive reviews including Dr. Grant Cates at The Aerospace Corporation and Dr. Paul Beery at Naval Postgraduate School. Their feedback was instrumental in improving our book.

The CRC Press staff was very helpful, especially Allison Shatkin for her great assistance and patience with us. We were also supported very much by our families letting us work on the book while taking time away.

List of Figures

List of Tables

1 Introduction

1.1 BOOK SCOPE

The title of this book indicates that it is directed toward engineers, although it is actually suited for a broader audience of engineers, scientists, and managers of technical organizations. These are people who perform or manage technical work but have not studied modeling and simulation (M&S).

These days, virtually all complex engineering projects involve some M&S. The increasing use of simulation modeling has been enabled by the growth of computing and is expected to continue. Consequently, more people must be conversant in the concepts of M&S.

Some M&S concepts are generally not well understood. These include the role of simulation in human learning and problem solving, representations of uncertainty in modeling, and degrees of necessity for verification and validation. This book addresses these concepts to provide a basic understanding of simulation modeling for those who have not studied the discipline.

This book is also intended for those who would like to learn simulation modeling with a general purpose simulation software program. While this book cannot provide thorough treatment of any particular simulation program, it does seek to make general purpose simulation programs more accessible to the beginner in simulation. To this end, several general purpose simulation programs are shown and explained to draw out the general methods underlying each simulation paradigm.

1.2 SYSTEMS AND MODELS

A *system* is a set of interacting parts that form a connected whole. A more thorough definition is:

> A *system* is a construct or collection of different elements that together produce results not obtainable by the elements alone. The elements, or parts, can include people, hardware, software, facilities, policies, and documents; that is, all things required to produce systems-level results. The results include system level qualities, properties, characteristics, functions, behavior and performance. The value added by the system as a whole, beyond that contributed independently by the parts, is primarily created by the relationship among the parts; that is, how they are interconnected [14].

A system must be represented in some form in order to analyze it and communicate about it. A *model* in the broadest sense is a representation of reality, ranging from physical mockups to graphical descriptions to abstract symbolic models. A model in the context of this book is a logical, quantitative description of how a system or process behaves. Though both *engineered* systems and natural systems are modeled in simulations, this book focuses on human-designed engineered systems. The models are abstractions of real or conceptual systems used as surrogates for low-cost experimentation and study. Models allow us to understand a system by dividing it into parts and looking at how they are related.

1.3 WHAT IS SIMULATION MODELING AND WHY MODEL?

Engineers design systems with many different kinds of models including mental models, diagrams and schematics, scaled physical models, mathematical models, and computational models. All are representations that enable understanding, design, and building of processes or systems. Many models are static representations where time plays no role, such as a scale model of a building, an equation for estimating the cost of an airplane, or a Bayesian network of contributing factors to an industrial accident. This book discusses modeling for simulation, which usually, but not always, implies dynamic modeling of process or system changes over time.

Simulation is the numerical evaluation of a mathematical model describing a system of interest. Many systems are too complex for closed-form analytical solutions, hence simulation is used to exercise models with given inputs to see how the system performs. Simulation can be used to explain system behavior, improve existing systems, or to design new systems too complex to be analyzed by spreadsheets or flowcharts.

A simulation model is a representation of a system that consists of variables related through formulas. Simulation software provides a means of specifying the structure of a model through formulas and performs the task of automatically updating variable values. At any particular moment in a time-based simulation, a model's state consists of its structure and the values of its variables.

Engineered systems are often too complex to represent in a closed form solution, an equation that solves a given problem in terms of functions and mathematical operations. For complex systems where analytical solutions aren't possible, one can use a simulation model consisting of linked equations that are executed whereby the outputs of some are inputs to others.

Simulation models are used for many reasons. One of these is to understand how a process or system works, particularly when inputs may vary and a model's structure contains feedback loops and delays. The effects of such complexity can be very difficult to understand without a model that allows one to represent and trace behavior.

Simulation modeling can play an important role in training and learning. Flight simulators are routinely used to train pilots to reduce risk by practicing in scenarios before actual flying. Simulators have also been used to teach project management decision-making skills, such as how much a schedule can be reduced before product quality is undermined.

Another reason for simulation modeling is supporting system or process design or redesign. Modeling can be used to compare designs, identify trade-offs, and choose among alternatives that provide the best support for the desired outcomes. Along with design, modeling can support planning the use of resources and forecasting outcomes using objective criteria.

When designing or planning, the cost of simulation modeling may appear high. At those times, one must determine how much is unknown about a system or process and then ask whether the cost of learning is higher with a model or with a trial implementation. One will often find that a program cannot afford not to model.

1.4 OVERVIEW OF SIMULATION MODEL TYPES

Static models do not involve time in their computations. There is no change of variable values over time. *Dynamic* models are used to simulate the time-varying behavior of a system. In time-based simulation the system is considered a collection of interacting elements, the properties of which change with time. These elements are typically referred to as *entities* and their properties as *attributes*. The interactions are often called *activities*. Continuous and discrete simulation models must be distinguished however. The decision as to which type of simulation model to use depends upon the nature of the system and the purpose of the simulation study (see Chapters 2 and 3).

Continuous models compute differential or algebraic equations at regular intervals to update state variable values to represent continual change over time. A continuous model shows how values of the attributes change as functions of time, computed at equidistant small time steps. These changes are typically represented by smooth continuous curves. Continuous models are useful in such areas as engineering design where well-established mathematical relationships give rise to models consisting of differential or algebraic equations. An example is computing the level of water in a dam based on continuous flow rates. Other areas of application address long-range trends, or growth and decay behavior. In these studies, many variables can be conveniently aggregated and local discrete fluctuations ignored. Continuous simulation models may also be applied to systems that are discrete in real life but where reasonably accurate solutions can be obtained by averaging values of the model variables. Examples of such approximations are traffic pattern studies or analysis of mass production by assembly lines. In these cases, one is not interested in individual cars or machine parts, but desires to know general traffic trends or the overall efficiency of a manufacturing process.

Continuous models solving differential equations are not all time-based. Integration may be across spatial distances (e.g., modeling thermodynamics or material stresses) instead of time progression. The underlying modeling procedures and numerical methods are structurally similar to time-based models.

Discrete or *discrete event* models generally consist of entities and their attributes that change during the simulation runs at event times. These changes occur instantaneously as the simulated time lapses, and are reflected in the output as discontinuous fluctuations. There are many systems that should be modeled as discrete because no continuous approximations are valid. An example situation is a model built to study the effects of different machine maintenance scheduling policies. In such a model one is concerned with queue lengths, waiting times, sequencing of related jobs, and availability of maintenance resources where one cannot ignore the discrete character of the system by aggregation or averaging.

Agent-based models can be considered forms of discrete models. They also model individual entities and use triggers to cause events. Agent-based models are distinguished by their encapsulation of behavioral rules within individual agents. Agents operate autonomously, interacting with an environment and with each other. The states of agents can change with an event like entity characteristics change in discrete event modeling. A similarity with continuous systems is that transitions move agents between states like continuous flows move entities between levels (continuous state variables). Agents may interact in discrete, continous, or combined environments.

A simulation model can be *deterministic*, *stochastic*, or *hybrid* (mixed). In the deterministic case, all input parameters are specified as single values. Stochastic modeling recognizes the inherent uncertainty in many parameters and relationships using random numbers drawn from a specified probability distributions. Mixed modeling employs both deterministic and stochastic parameters.

In a purely deterministic model, only one simulation run is needed for a given set of parameters. However, with stochastic or mixed modeling the result variables differ from one run to another because the random numbers drawn differ from run to run. In this case the result variables are best analyzed statistically (e.g., mean, standard deviation, distribution type) across a batch of simulation runs. This is termed *Monte Carlo simulation* or Monte Carlo analysis (Chapter 4).

1.5 APPLICATIONS

To illustrate the broad spectrum in which M&S can be used, representative engineering applications are shown in Table 1.1 for general engineering disciplines. The list is not exhaustive across disciplines, inclusive of all applications, or static. Not shown are many subfields or specialties such as aerospace engineering, automotive engineering, environmental engineering, engineering management, robotic engineering, manufacturing engineering, and countless others. These examples are common traditional applications and more areas will come into vogue as the fields advance. The applications also have subcategories not shown. A good number of applications are shared across disciplines (e.g., materials science in mechanical and/or chemical engineering) so the traditional closest fits are shown.

TABLE 1.1: Representative Engineering Applications

Discipline	Applications
Biomedical	Biomechanics
	Biomedicine
	Medical devices
	Bioinformatics
	Biomaterials

(*continued*)

TABLE 1.1: Representative Engineering Applications (*continued*)

Discipline	**Applications**
Chemical	Biochemical
	Corrosion
	Genetics
	Pharmaceuticals
	Transport phenomena
	Waste management
	Plant design
	Process design
	Chemical synthesis
Civil	Structural design
	Building materials
	Hydraulics
	Transportation systems
	Environmental
	Municipal infrastructure
	Construction management
Electrical and Electronic	Circuit analysis and design
	Power systems and grids
	Signal processing and communications
	Networks
	Computer engineering
Mechanical	Materials
	Thermodynamics
	Fluid dynamics
	Vehicle dynamics
	Kinematics
	Nanotechnology
	Control system design
Software	Software development processes
	Software evolution and maintenance
	Defect modeling
	Business value
Industrial and Systems	Systems of Systems
	Sustainment
	System Architecture

(*continued*)

TABLE 1.1: Representative Engineering Applications (*continued*)

Discipline	Applications
	Manufacturing
	Operations research
	Economic and financial analysis

1.6 INTRODUCTORY EXAMPLE WITH BASIC MEASURES

The methods of modeling and simulation will be demonstrated first with a running example for electric car charging scenarios starting with a simple scenario that can be manually calculated. In subsequent chapters we will extend it with increasingly complex working models and simulations to demonstrate the expanding concepts.

The case study example is for a company that develops, manufactures, and installs electric car charging station equipment. The charging stations are for customers "on-the-go" for short-term charges. The company needs to decide where to install charging stations, how many charging bays to include at each location, and what options for charging power capability need to be provided. They cannot address these questions with a simple mathematical solution because of the complexity of operations and system uncertainties, thus they will simulate the operations to support the decisions.

To gain insight they will analyze standard metrics from the simulations including waiting time, queue length, and resource utilization. This is because excessive wait time leads to unhappy customers, drives customers to competitors' services, and reduces profit. Queues require physical space. If the queue cannot accommodate enough cars safely, then cars won't even be able to enter the queue. If queues are too long, customers may also leave. For resource utilization, underused resources are wasteful and reduce profit. An overused resource indicates that another resource may be warranted, and it leads to excessive waits and queue lengths. Optimizing the waiting time, queue length, and resource utilization will be key to success or failure.

1.6.1 QUEUE-SERVER SYSTEM

The most common discrete system is the queue/server system. Queue/server systems exist whenever a line forms in front of some processing mechanism. The line is called a *queue* and the processing mechanism is called the *server*. In this example the cars form in queues to be served by the charging bay resource. Modeling the scenario entails quantifying the car arrival times and their respective charging (service) times.

In this first manual example we'll assume some random values for car arrivals and charging times. For simplicity we'll use integer values for easy calculation and assess the first one hour of operations using basic measures for simulation output shown in Table 1.2. These are the per-car calculations.

Assume the first eight cars arrive at a single charging bay with the following inter-arrival times in minutes (denoting the times between successive arrivals) starting at $time = 0$:

$$[2, 8, 7, 2, 11, 3, 15, 9].$$

Their respective charging times in minutes at the station are

$$[11, 8, 5, 8, 8, 5, 10, 12].$$

These are the inputs to compute through a simple discrete event simulation as illustrated.

For the car charging discrete event model we'll compute basic measures to quantify the charging service operations. We will calculate the minimum and maximum queue length, number of customers served at the one-hour mark, the total idle time for the charging bay, the total queue time, and the average waiting time in queue. Table 1.2 shows the resulting events and measures using the given arrival and charging times.

TABLE 1.2
Example Discrete Event Calculations

Car #	Inter-arrival Time	Charging Time	Clock Time	Charging Start	Charging End	Wait Time	Charger Idle Time	Queue Length Start	Queue Length End
1	2	11	2	2	13	0	2	0	1
2	8	8	10	13	21	3	0	1	2
3	7	5	17	21	26	4	0	2	1
4	2	8	19	26	34	7	0	1	2
5	11	8	30	34	42	4	0	2	2
6	3	5	33	42	47	9	0	2	0
7	15	10	48	48	58	0	1	0	1
8	9	12	57	58	70	1	0	1	0

Table 1.3 shows system snapshots of all the events and statistics simulating this scenario as time progresses. It includes the event list and updated statistics at each event. The upcoming event list consists of arrival and departure events, of which the closest one in time is scheduled next for processing. The clock is updated at each event invocation and the ongoing statistics re-calculated. Additional mathematical description of this next event time advance approach is in Section 3.2.1.

Resource Utilization

An important measure of system performance is a measure of how busy the server is. The utilization of the server is the proportion of time during a simulation that the server is busy (not idle). The quantity can be measured as a continuous time average (similar to average queue length) by defining a binary busy function:

$$B(t) = \begin{cases} 0, \text{ if } x = 1 \\ 1, \text{ if } x = 2. \end{cases} \tag{1.1}$$

The utilization $u(n)$ is the proportion of time that $B(t)$ is equal to 1 as in Equation 1.2. In the equation, $u(n)$ is the continuous average of the $B(t)$ function, which matches intuition about utilization.

TABLE 1.3: System Snapshots with Events and Statistics

Clock	Server	Queue	Statistics
00:00			Cars Entered = 0 Cars Served = 0 Total Time in Queue = 0
Initialization	**Next Arrival**	**Next Departure**	Mean Waiting Time = N/A
	Car 1 at t=2	-	Utilization = N/A

Clock	Server	Queue	Statistics
00:02			Cars Entered = 1 Cars Served = 0 Total Time in Queue = 0
Event	**Next Arrival**	**Next Departure**	Mean Waiting Time = 0
Car 1 Arrival	Car 2 at t=10	Car 1 at t=13	Utilization = 0

Clock	Server	Queue	Statistics
00:10			Cars Entered = 2 Cars Served = 0 Total Time in Queue = 0
Event	**Next Arrival**	**Next Departure**	Mean Waiting Time = 0
Car 2 Arrival	Car 3 at t=17	Car 1 at t=13	Utilization = 0.8

Clock	Server	Queue	Statistics
00:13			Cars Entered = 2 Cars Served = 1 Total Time in Queue = 3
Event	**Next Arrival**	**Next Departure**	Mean Waiting Time = 1.5
Car 1 Departure	Car 3 at t=17	Car 2 at t=21	Utilization = 0.85

(*continued*)

TABLE 1.3: System Snapshots with Events and Statistics (*continued*)

Clock	**Server**	**Queue**	**Statistics**
00:17			Cars Entered = 3 Cars Served = 1 Total Time in Queue = 3
Event	**Next Arrival**	**Next Departure**	Mean Waiting Time = 1.5
Car 3 Arrival	Car 4 at t=19	Car 2 at t=21	Utilization = 0.88

Clock	**Server**	**Queue**	**Statistics**
00:19			Cars Entered = 4 Cars Served = 1 Total Time in Queue = 3
Event	**Next Arrival**	**Next Departure**	Mean Waiting Time = 1.5
Car 4 Arrival	Car 5 at t=30	Car 2 at t=21	Utilization = 0.89

Clock	**Server**	**Queue**	**Statistics**
00:21			Cars Entered = 4 Cars Served = 2 Total Time in Queue = 7
Event	**Next Arrival**	**Next Departure**	Mean Waiting Time = 2.3
Car 2 Departure	Car 5 at t=30	Car 3 at t=26	Utilization = 0.90

Clock	**Server**	**Queue**	**Statistics**
00:26			Cars Entered = 4 Cars Served = 3 Total Time in Queue = 14
Event	**Next Arrival**	**Next Departure**	Mean Waiting Time = 3.5
Car 3 Departure	Car 5 at t=30	Car 4 at t=34	Utilization = 0.92

Clock	**Server**	**Queue**	**Statistics**
00:30			Cars Entered = 5 Cars Served = 3 Total Time in Queue = 14
Event	**Next Arrival**	**Next Departure**	Mean Waiting Time = 3.5
Car 5 Arriving	Car 6 at t=33	Car 4 at t=34	Utilization = 0.93

(*continued*)

TABLE 1.3: System Snapshots with Events and Statistics (*continued*)

Clock	**Server**	**Queue**	**Statistics**
00:33			Cars Entered = 6 Cars Served = 3 Total Time in Queue = 14
Event	**Next Arrival**	**Next Departure**	Mean Waiting Time = 3.5
Car 6 Arrival	Car 7 at t=48	Car 4 at t=34	Utilization = 0.94

Clock	**Server**	**Queue**	**Statistics**
00:34			Cars Entered = 6 Cars Served = 4 Total Time in Queue = 18
Event	**Next Arrival**	**Next Departure**	Mean Waiting Time = 3.6
Car 4 Departure	Car 7 at t=48	Car 5 at t=42	Utilization = 0.94

Clock	**Server**	**Queue**	**Statistics**
00:42			Cars Entered = 6 Cars Served = 5 Total Time in Queue = 18
Event	**Next Arrival**	**Next Departure**	Mean Waiting Time = 3.6
Car 5 Departure	Car 7 at t=48	Car 6 at t=47	Utilization = 0.95

Clock	**Server**	**Queue**	**Statistics**
00:47			Cars Entered = 6 Cars Served = 6 Total Time in Queue = 27
Event	**Next Arrival**	**Next Departure**	Mean Waiting Time = 4.5
Car 6 Departure	Car 7 at t=48	-	Utilization = 0.96

Clock	**Server**	**Queue**	**Statistics**
00:48			Cars Entered = 7 Cars Served = 6 Total Time in Queue = 27
Event	**Next Arrival**	**Next Departure**	Mean Waiting Time = 4.5
Car 7 Arrival	Car 8 at t=57	Car 7 at t=58	Utilization = 0.94

(*continued*)

TABLE 1.3: System Snapshots with Events and Statistics (*continued*)

Clock	Server	Queue	Statistics
00:57			Cars Entered = 8 Cars Served = 6 Total Time in Queue = 27
Event	**Next Arrival**	**Next Departure**	Mean Waiting Time = 4.5
Car 8 Arrival	-	Car 7 at t=58	Utilization = 0.95

Clock	Server	Queue	Statistics
00:58			Cars Entered = 8 Cars Served = 7 Total Time in Queue = 27
Event	**Next Arrival**	**Next Departure**	Mean Waiting Time = 3.86
Car 7 Departure	-	Car 8 at t=70	Utilization = 0.95

$$u(n) = \frac{\int_0^{T(n)} B(t)}{T(n)} \tag{1.2}$$

Waiting Time

The basic statistics for waiting time are the mean (average) waiting time in Equation 1.3 and sample variance in Equation 1.4.

$$\overline{W} = \frac{1}{n}\sum_{i=1}^{n} W_i \tag{1.3}$$

$$Var_W = \frac{1}{n-1}\sum_{i=1}^{n} (W_i - \overline{W})^2 \tag{1.4}$$

The mean waiting time for the first hour of operation (seven cars completing) is

Mean Waiting Time ($\overline{W}$) = (0 + 3 + 4 + 7 + 4 + 9 + 0) / 7
= 3.86 minutes.

Queue Length

The mean and variance of the queue length are in Equations 1.5 and 1.6 where T is the total simulation time and t_i is the length of time between the $(i-1)^{\text{th}}$ and i^{th} events. A graph of queue length is shown in Figure 1.1 and continued in Figure 1.2 (which can also be derived from Table 1.2).

$$\overline{L} = \frac{1}{T}\sum_{i=1}^{n} L_i t_i \tag{1.5}$$

$$Var_L = \frac{1}{T}\sum_{i=1}^{n} (L_i - \overline{L})^2 t_i \tag{1.6}$$

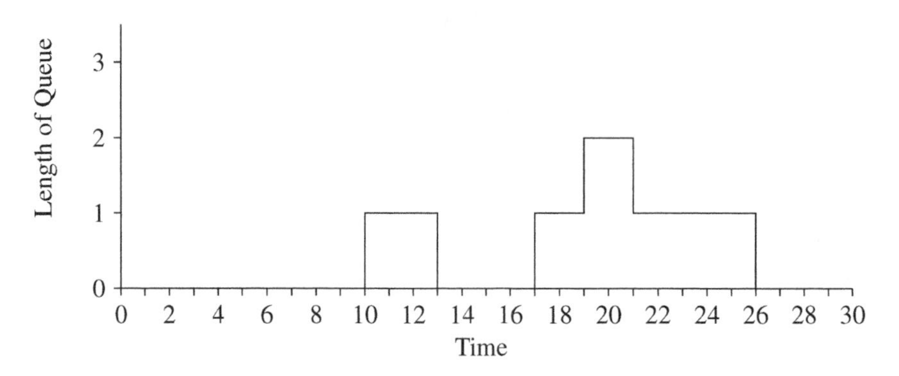

FIGURE 1.1: Length of Queue (0–30 Minutes)

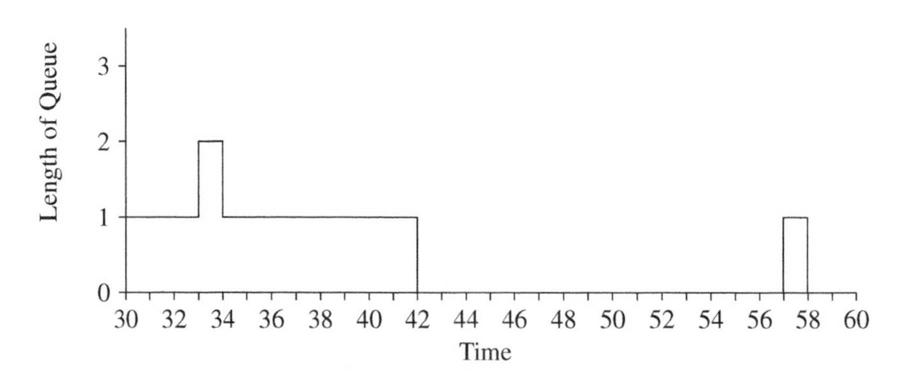

FIGURE 1.2: Length of Queue (30–60 Minutes)

$$\begin{aligned}\text{Average queue length } (\overline{L}) &= [(10-0)0+(13-10)1\\ &\quad +(17-13)0+(19-17)1+(21-19)2+(26-21)1\\ &\quad +(30-26)0+(33-30)1+(34-33)2+(42-34)1\\ &\quad +(57-42)0+(58-57)1+(60-58)0]/60\\ &= [0+3+0+2+4+5+0+3+2+8+0+1+0]/60\\ &= .47 \text{ cars}\end{aligned}$$

A graph of the utilization for the charging resource is shown in Figure 1.3 and continued in Figure 1.4 (which can also be derived from Table 1.2).

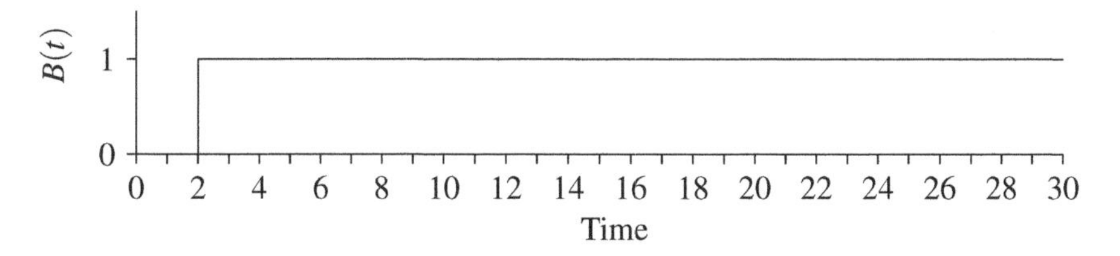

FIGURE 1.3: Resource Utilization (0–30 Minutes)

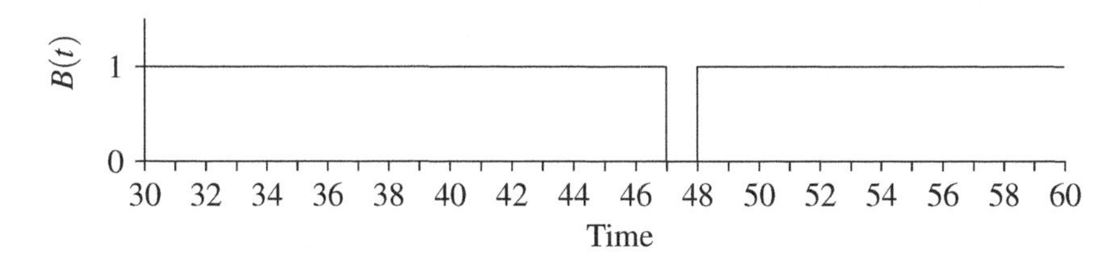

FIGURE 1.4: Resource Utilization (30–60 Minutes)

The resource utilization of the charging server for the simple example would be calculated as:

$$\begin{aligned} \text{Utilization } u(n) &= [(2-0)0+(47-2)1 \\ &\quad +(48-47)0+(60-48)1]/60 \\ &= [0+45+0+12]/60 \\ &= .95. \end{aligned}$$

Thus the server is busy 95% of the time during the 60 minutes of simulation time. This car charging station example is the basis for further elaboration in subsequent chapters to illustrate the modeling process and concepts. It will be constructed as a stochastic model with randomness, inputs will be varied, it will run with various scenarios, outputs will be tested statistically, and conclusions drawn from the simulation experiments.

1.7 SUMMARY

Engineers, scientists, and managers should be well informed about M&S concepts. Virtually all complex engineering projects involve some modeling and simulation. Purely analytic methods seldom apply to complex, engineered systems with inherent uncertainties. Thus simulation models are used to understand how a system or process works, to design or redesign an engineering process, and to support training and learning.

A system is defined as a set of interacting parts that form a connected whole producing results not achievable by the elements alone. It must be represented in some form in order to analyze it and communicate about it. A model in this book's context is a logical, quantitative description of how a system or process behaves. Simulation is the numerical evaluation of such a mathematical model describing a system of interest.

The models are abstractions of real or conceptual systems used as surrogates for low-cost experimentation and study. Models allow us to understand a system or process by dividing it into parts and looking at how they are related. This book focuses on dynamic models used to simulate the time-varying behavior of a system. In time-based simulation the system is considered a collection of interacting elements, the properties of which change with time. These elements are referred to as entities and their properties as attributes.

The most popular types of dynamic simulation models are continuous, discrete event and agent based. A continuous model shows how values of the attributes change continuously as functions of time computed at equidistant small time steps. These changes are typically represented by smooth continuous curves. Continuous models may also be applied to systems that are actually discrete when the outputs have sufficient accuracy.

Discrete or discrete event models consist of entities and their attributes that change during simulation runs at aperiodic event times. These changes occur instantaneously as the simulated time lapses, and are reflected in the output as discontinuous fluctuations.

Agent-based models share common aspects with discrete models. They also model individual entities and use triggers to cause events, but the entities are autonomous agents that encapsulate their own rules of behavior. They operate within an external environment and interact with each other. At event times they may change their internal state.

Furthermore, simulation models can be deterministic, stochastic, or a hybrid. Most systems and models of concern are stochastic whereby the result variables differ from one run to another because of inherent randomness. For this the results must be analyzed statistically across multiple simulation runs to determine probabilistic means, standard deviations, distributions, etc. Monte Carlo simulation is commonly used to run models many times with random inputs for this purpose.

There is a very broad spectrum of applications in which modeling and simulation can be used across all engineering disciplines. In all these areas, simulation models are ultimately used to support engineering decision making based on the simulation results.

The most common discrete system is a queue/server system. In these systems, entities form lines (queues) in front of processing mechanisms called servers that provide resources. The entities use the resources to perform activities in the system. To gain insight for decisions, standard metrics from discrete event simulations include waiting time, queue length, and resource utilization. These measurements have impact in real-world operations and thus answer questions to address modeling goals.

2 Modeling and Simulation Method

2.1 INTRODUCTION

Modeling and simulation (M&S) projects vary widely according to the nature of what is represented, the simulation paradigm employed, the modeling purpose, and the size of the modeling effort. Therefore, modeling processes and methodologies employed can be very different across projects. For example, a continuous model of traffic patterns might be constructed within a few days by a single modeler to illustrate the effects of a road maintenance project over a two-year period. On the other hand, a large discrete event model for managing automobile manufacturing supply chains might be developed and maintained by a large modeling team.

Simulation is used to represent the behavior of systems, processes, or scenarios. A system view is concerned with inputs to a system boundary, their transformation, and outputs across the system boundary. A process view is concerned with a series of steps, each step having inputs, performing a transformation, and producing an output. Scenarios in the context of training or gaming describe an environment with situational factors and actions performed. This book is primarily concerned with modeling and simulating real systems and processes, though it is not meant to exclude scenarios. In this book, a reality represented in a model may be referred to as a system, a system or process, a subject target, or a subject.

Understanding behavior is the usual reason for M&S. Even a simple process can have complex behavior, but simulation models often become complex because they represent systems or processes composed of many elements having many relationships and the overall behaviors are complex. Real complexity can complicate the modeling unnecessarily unless a disciplined approach is used.

The basis for a disciplined approach is the sequence of modeling stages discussed in this chapter with the caveat that modeling is not a strictly sequential process of engagement, specification, construction, verification and validation, experimentation, and reporting shown in Figure 2.1. Rather, a modeler undertakes these stages iteratively. For example, during engagement, a modeler may realize that a particular part of a system would be most difficult to represent, and so may specify that part at a high level of abstraction for discussion with the sponsor. The modeler may even construct a small model to take back to the sponsor for elucidation of the system workings and validation of modeling scope. Elaboration of the model would proceed with further iterations that respect the sequence of the stages.

Despite the wide variety of modeling projects, a general method for modeling and simulation can be described. As it is described, the need for specialized knowledge and skills will become apparent, therefore these are discussed later in this chapter.

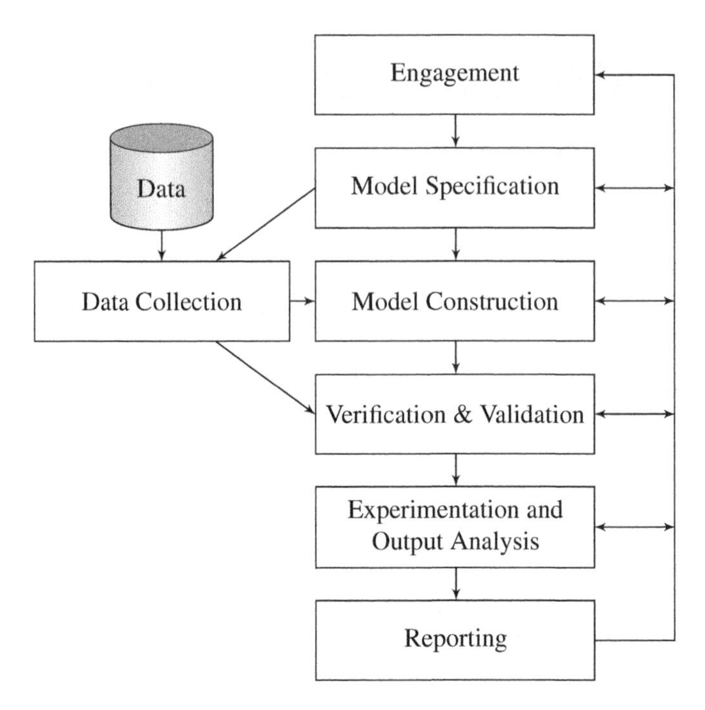

FIGURE 2.1: Modeling Process

2.2 ENGAGEMENT

When engaging with a customer for a modeling project, the customer may not have been exposed to M&S studies and so may not know how an M&S study can help. In this case, the modeler must spend time with the customer discussing what problems the customer perceives, why the customer thinks an M&S study may help, what can be represented, and how an M&S study can actually help. The customer is also likely to want to know about the modeler's experience and the cost and benefits of previous studies. If the customer is familiar with M&S studies, then the customer may have realistic expectations of the possibilities and the initial conversation can proceed directly to a particular problem, which must be important enough to the customer to warrant an M&S study.

Once a modeler and customer decide to engage in a modeling study, several points are important. The first of these is identifying the stakeholders, including sponsors (those who manage the funding), process or system owners (those responsible for managing the real system or process and the people in it), and the process or system participants (the people who do the work). Beyond a cognizant engineering specialist, stakeholders may include other engineers and scientists in related disciplines using the simulation results, managers and executives deciding on product options, providers of input data, system or product testing, marketing and sales, finance, human resources, investors, government and regulatory agencies, or others impacted depending on the modeling context.

Each stakeholder may have a different perspective on the reality to be represented and the modeler's task is to reconcile their viewpoints using a model. The modeler will have to facilitate discussions that elicit information about the reality in order to represent its structure, its inputs, and its expected and actual outputs. The modeler must also negotiate a model purpose, project budget and schedule, access to each group of stakeholders, access to data, and a model specification including model scope and level of abstraction.

In cases where a modeler has extensive experience in the domain that he/she is modeling, then he/she may not have to rely heavily on domain experts. Usually, though, a modeler who is engaging with a customer must draw heavily upon the experts in the customer organization. The best experts for an M&S study can be those who directly manage the people participating in a system or process because they are the most informed. In an engineering organization, these experts are often technical leads or first-level managers.

Beyond initial sessions in which the modeler elicits information about the system or process, the modeler must ensure that access is available for follow-up discussion of questions that arise during model construction, verification and validation, and experimentation. At the same time, the modeler must keep stakeholders informed of the study and its progress, mindful that stakeholders unfamiliar with M&S may not appreciate the complexity of their own systems and challenges in representing them.

The preceding discussion can be summarized in the following list of questions that must be answered when engaging for an M&S project or study:

Who are the (groups of) stakeholders and what is the primary interest of each with regard to the modeling project?
What are the roles of those involved in the project?
Who will ensure that resources are available for the project or study?
Who needs to be kept informed of progress and issues?
Who will the modeler depend on for providing information and quantitative data about the real system or process? What is their availability?
What sources of data are available?
How much time is available for the M&S project or study?
What budget is available for the M&S project or study?

2.3 MODEL SPECIFICATION

Model specification usually begins with identifying a problem and describing its context as a basis for specifying a modeling purpose. The modeling purpose will include the intended audience for the modeling results, the desired set of policies to be simulated, and the kinds of decisions the results are intended to support.

Modeling purposes can be characterized in two dimensions, utility and generality. Three categories of utility are distinguished here.

1. A model can provide an object for discussion as a concept is being defined and social constructs are developed. Conceptual definition models may not go through all the modeling stages, particularly verification and validation, because they serve as conceptual illustrations and support identification of variables that can strongly influence system behavior. Models developed for this first purpose have a very limited life, as they can be thrown away once they have served their purpose, or they may become a basis for more extensive modeling.
2. Models are also built to address a specific problem or question. For example, an organization may want to know how many servers should be specified for a system designed to handle an expected volume of transactions. A model of this kind may be produced for a specific system development project and its usefulness ends with the project, though it can be archived for a later time when a similar project might be undertaken.
3. Models may be developed and maintained for long-term usage. Flight simulators fall into this category. Also in this category are models that represent operations and provide ongoing decision support for operational planning.

In specifying the purpose of a model in the first two categories, it is helpful to pose the single most important question that the model will be used to address. It is also very helpful to establish the graphic that will be used to answer the question, the single most important chart that the customer wants to see. For example, it could be a cumulative distribution function for the time to complete a project, or a quantity distribution function for the number of widgets produced. Whatever it is, it needs to be specified early so that the simulation can be made to produce that chart.

The model may answer questions beyond the most important one originally posed for the model, but the one question helps to focus the model purpose and provides a criterion for model scope decisions. Though a model may have a single, well-defined purpose, it may well provide information on related issues. For example, a model constructed to answer a question about the duration of a process may also indicate what resources are constraining the process. Chapter 7 relates an example illustrating this point.

Modeling purposes in the third category lead to large endeavors that require experience gained from conceptual and limited use models. These types of modeling projects can address multiple purposes difficult to express in a single question. In these cases of larger utility, managing model purpose, scope, and level of abstraction require considerably more coordination and discipline.

Modeling purposes may also be distinguished by their degree of specificity or generality. A model may be built to represent a specific system or process, or a class of systems or processes. General models have much larger audiences than specific models. Very specific models require data from only one source, but general models require data from many sources. Verifying and validating specific models can also require less effort than general models. When specifying the purpose of a model, a modeler should keep in mind that more work comes with greater utility and with greater generality.

The primary modeling purpose question is:

What question(s) should the modeling results answer?

Model scope may be defined by the modeling purpose, but it is clarified by the boundaries defined for the model. This means that the model must include elements sufficient to generate the outputs of interest. For example, if a customer is interested in resource utilization, then all the demands on the resources should be included. Modeler and stakeholders have to agree on what is included and excluded from the model. Maintaining model scope can be challenging. Stakeholders will have expectations that may go beyond the model scope, so the modeler must remind them of scope and capability. Model scope and boundaries can grow from one model version to the next, but scope should be maintained and clearly communicated for each version.

Modelers face a temptation against good definition. It is the desire to satisfy all stakeholders by providing a model that represents all their perspectives rather than a model that represents a common concern. For example, one stakeholder may want a model that reflects resource usage for a process in order to determine what the capacity of the process is, but another may want a model that includes all the personnel in their organization so that they can see where all their labor is spent, including that spent on the first stakeholder's process of interest. In that case, the modeler may need to use his negotiation skills to facilitate an agreement among the stakeholders for the sake of producing an adequate model within budget and schedule constraints.

The following are model scoping questions:

Where should the model "start"? Where should it "end"?
What system elements should be included?
What resources are to be included?
For each model component, will inclusion significantly affect the results? If so, will the influence on results be optimistic or pessimistic?
For each model component, will inclusion strongly affect the model's credibility?
For each model component, is data available or obtainable? How accurate is it?

Every model is an abstraction of reality and the modeler must choose an appropriate "level" of abstraction for a model. Choosing levels of abstraction can be challenging for a modeler, but it also allows the modeler to control how much detail goes into a model. A good guideline is to model at the level of abstraction necessary to answer the questions. Simple is typically better than complicated. When in doubt, by default start with as simple a model as possible and add detail later as necessary. One technique used successfully by expert modelers is to start with a simple prototype or proof-of-concept model early in the project to work out potential issues with the project, to improve the accuracy of the project time and resource estimates, and to obtain support from the stakeholders.

As an example of level of the abstraction, a model that simulates satellite communications can be very detailed if used to design communication scheduling algorithms, but can be very high level if used to confirm a decision on the number of satellites required in an orbit. The choice of abstraction level is also qualified by an M&S project budget and schedule. A modeler working within a small budget and tight schedule can choose to produce a smaller, more abstract model that requires less work to build and verify than a large, high-fidelity model. In fact, this approach is recommended: if the customer sees that results of a small model are useful, then addition of details for a better fidelity can be negotiated.

The following are model abstraction questions:

Which of the following factors are influencing the modeling level of abstraction: modeling objective, model scope, performance measures of interest, alternatives to be examined, examination details for alternatives, data availability, execution time, modeling budget and schedule, animation, modeling toolset and language support, and modeler skillset? Is the degree of influence of each factor commensurate with the modeling purpose?
How should each element be represented and to what level of detail?
What simplifying assumptions can be made?
Will any of the assumptions bias the modeling results either pessimistically or optimistically?
What alternatives are to be explored?
What scenarios are to be considered?

As a model is specified, important model parameters are discussed. These discussions typically start with output variables that address the modeling purpose and input variables that domain experts recognize as influential. With the discussion of input parameters, sources of quantitative data are discussed. Obtaining data for model parameters can be the most time-consuming part of a modeling project for several reasons. Data for model parameters may not exist and it will need to be elicited or collected. If data does exist, it may be proprietary and 1) upper management will need to be convinced that it should be made available for the modeling project, and 2) a process for acquiring the data will need to be executed. When data is provided, reviewing it and scrubbing it may require substantial effort.

Due to potentially long lead times in obtaining data, discussion of data sources should be undertaken during model specification so that data can be available once a model is constructed. The modeler requires data for model parameters but also for actual outcomes of the subject reality for model validation. Setting up a data collection system may not be feasible within the schedule and budget of the modeling project. If that is the case, then data is limited to whatever is available or whatever estimates can be elicited from domain experts.

The following are data acquisition questions:

What sources of data are available for model inputs?
If data is lacking for expected model inputs, can estimates be elicited from domain experts?

What system performance data can be provided for validating the model?

When modeling a specific system or process, one might attempt to learn the real system or process through documents and telephone interviews. However, trying to model from a distance is difficult and inhibits fair representation of the reality. The best opportunity for good modeling occurs when making site visits, meeting the stakeholders in person, and walking through the subject of the model.

The foregoing discussion can be summarized in the following questions that should be answered when specifying a model:

What is the question to be answered by a model study?
What is the intended utility of the model: conceptual definition, solution generation, or long-term support?
Is the model to represent a specific case or a class of cases?
What kinds of decisions should be supported by the study?
Who is the intended audience?
What set of policies should be modeled?
What is the model scope?
What is the model level of abstraction?
What model outputs are necessary?
What kind of output data analysis will be necessary?

2.4 MODELING

2.4.1 MODELING PARADIGMS

One of the first decisions to be made in modeling is choosing the modeling paradigm. This may be dictated by the modeler's expertise in a particular paradigm or the paradigms supported by a simulation software package used in the modeler's organization. However, the different paradigms have advantages and limitations for various types of modeling. Continuous modeling, discrete event simulation, and agent-based simulation are discussed here.

Continuous simulation and discrete event simulation are distinguished by their representation of time. In continuous simulation, time is represented in regular intervals. Flows are represented as passing through a continuous model into and out of containers (also called stocks or levels). Flow rates can be increased and decreased based on the variable relationships. With each tick of the simulation clock in a continuous model, a fixed amount of simulated time passes and the values of all model variables, flow rates, and container contents are updated.

In a discrete event model, events are scheduled and the simulation clock moves to each event in the schedule (see the next event time advance approach in Section 3.2.1). Entities move through a model, obtaining resources, spending time in activities, and releasing their resources. Entities can have attributes so that each entity can have its own characteristics. Attribute values are used to set activity durations and to route entities through a model. Variable values are updated with each event.

Because they represent flows, continuous models tend to be more abstract than discrete event models. Where a continuous model treats work as a continuous flow, a discrete event model treats work as individual items passing through activities. Consequently, discrete event model diagrams can be easier to understand for audiences that think of entities being changed through activities performed on them. The ability to characterize entities in discrete event models also adds to representational capability.

Agent-based models represent interacting entities, or agents. Agents have their own characteristics and can initiate actions, communicate with one another, and react to one another. Like discrete event simulation, agent-based models generally treat time as a series of discrete events.

Table 2.1 summarizes characteristics of the three modeling paradigms in terms of system characteristics and modeling goals. Use continuous modeling, such as system dynamics modeling, if a global view is desired and aggregated entities are sufficient, whereby information on individual entities isn't necessary or possible to model. Discrete event modeling is useful for investigating system-level process behavior with visibility into individual entities. Agent-based modeling is also useful for disaggregated modeling whereby individual object behavior can be described and the modeler is interested in collective behavior that emerges from the interactions of the agents.

TABLE 2.1
Model Type Selection Criteria

Criteria	Continuous	Discrete Event	Agent Based
Perspective	Global view with feedback	System-level processes	Individual interacting objects
Entities	Aggregated entities	Disaggregated entities with attributes	Disaggregated agents with behaviors

Having acknowledged the distinctions for modeling paradigms, it is also important to know that current simulation programs are blurring these distinctions, particularly between continuous and discrete event simulation. The programs support both continuous and discrete event timing in the same model and is sometimes called hybrid modeling.

The choice of a modeling paradigm may depend on the simulation capabilities required as well as the modeler's expertise. These three modeling paradigms are discussed in more detail in Chapter 3.

2.4.2 MODEL CONSTRUCTION

Once a modeling paradigm and supporting software have been selected, decisions must be made about representing the real subject of the model. For continuous models, anything in the real system that changes is represented as a flow, so one must decide what flows to represent, how they are modified, and how they interact. For discrete event models, one must decide how to represent system elements using entities and resources, and how entities are routed. For agent-based models, agents are identified, their characteristics are specified and rules for their behavior are written.

In making decisions about model structure, multiple representational possibilities may be available. Consider, for example, modeling of a workflow process in a discrete event model. Entities may be used to represent work items and people are treated as resources, or entities may represent people who perform work and the work items are treated as resources. The choice of representation may depend on the modeling purpose. If the stakeholders are interested in changes in the people as they work, then the people can be characterized using attributes and attribute values that are updated to reflect changes in personal characteristics. On the other hand, if the stakeholders are interested in changes in work products and work capacity, then work items can be represented as entities with attributes, and people are the resources. When representational choices become apparent, simple trial models of each should be produced and evaluated.

Another structural decision in modeling is the division of the model into parts. Small models may not need partitioning, but larger models are managed best in parts. A number of factors may influence decisions about partitioning a model. The real system may logically be partitioned, suggesting the way in which a model of it should be partitioned. Also, the different flows or entity types or agents identified for a model may suggest partitions. Previous models of the subject reality may also suggest advantageous or disadvantageous ways to partition a model. In some cases, partitions are created to add new purpose and scope to an existing model.

As construction proceeds, a modeler should distinguish between design and implementation. For those who use graphical simulation software, the difference between design and implementation can be blurred because part of a model can be a working implementation while another part is being designed. A temptation faced by all modelers, especially novices, is to create an entire design of a large model in a graphical program with the expectation that the model runs as intended. However, the result can be an unwieldy model that is difficult to debug. When such a model has been created, it is recommended that it be treated as a design, set aside, and start a new model in which small sections can be implemented and tested, one section at a time. Of course, this iterative approach can and should be used from the outset, designing and implementing in pieces. An iterative approach with successive elaborations ("build a little, test a little") is best.

As a model is constructed, it should have only enough detail to address its purpose(s) and no more. Modelers face several temptations that can lead to models that are larger than necessary:

The desire to have the most comprehensive model in the field, one that represents all relevant phenomena known on the subject.
The desire to satisfy all stakeholders, who may be asking questions such as "Does it include . . . ?" and "Can it show . . . ?"
The desire to build and maintain only one model that incorporates all subsystems or incorporates all system variations.
The desire to avoid representational abstractions about which one is unsure. It can seem easier to add many details and be sure that the representation is correct rather than try more abstract representations that require less data and testing.

As these temptations arise, it is best to face each one and acknowledge the trade-offs that come. As models grow, so does the amount of work in constructing, testing, fixing, and managing them.

Depending on how a model is to be used, performance requirements may be necessary. Iterative construction also facilitates assessment of model performance as the model is being built. For example, if many thousands of runs are needed to produce the necessary output data, then the simulation software must be able to record all the runs without using all available computer memory. The model must also be able to run fast enough to produce the required number of runs in a reasonable time period. Performance requirements and desires should be documented and performance should be assessed periodically during model construction to determine whether computing resources are sufficient.

Iterative construction also facilitates configuration management of a model. As each addition is made to a model, the change can be documented either in the model file or a separate document. The model file can be copied to a new file and a new edition can be started. This management of configurations is especially helpful when experimenting with model constructs and wanting to branch from a previous version, when multiple modelers are working on the same model, while debugging and looking for the source of a problem, or when stakeholders are requiring results and a stable version is needed for producing them.

Modeling decisions can reflect trade-offs between multiplicity and behavioral complexity. For example, suppose a factory has multiple product assembly lines that have much in common but differ from one another in a few processing steps. A graphical model of these lines might depict each line separately or it might abstract from the separate lines by combining them into a single line with processing exceptions to represent the differences. Combining the lines produces a simpler static view of the model. However, all the items passing through one processing line can make tracing more difficult than tracing fewer entities through individual lines. In the end, one may decide to use a graphical depiction of multiple lines because the diagram is easier to use when discussing the model with stakeholders familiar with the factory layout.

2.4.3 DATA COLLECTION AND INPUT ANALYSIS

Parameter creation is part of model construction. Parameters for inputs and outputs should be readily identifiable to system managers and participants. Input parameters should be created with the expectation that data can be collected for them and output parameters should be created with the expectation that data from system outcomes is available for validation.

Data must be relevant and of good quality. Relevant data comes from measurements that have the same meaning and units as the model parameters for which the data will be used. Good quality implies that data is complete and does not contain anomalies, such as inaccurately recorded values. Obtaining good-quality, relevant data for input parameters and for validation can be challenging. Although some types of simulation models are built to use qualitative data, simulation models ordinarily use quantitative data.

Quantitative data can be deterministic or stochastic. Deterministic data can be either constant values or variable values computed as a function of time or other independent variables. Stochastic data reflects variation that cannot be characterized as a deterministic function. The uncertainty in values can be characterized in a random distribution. When recorded data is provided for input parameters, it can be analyzed to determine whether it should be used in a model as constant, functionally variable, or randomly variable. However, when recorded data is not available for input parameters, the modeler faces a question as to whether data should be collected or elicited.

If data collection is undertaken, the costs of developing a measurement system and collecting data are incurred. These include specifying the data to be collected, specifying the collection methods and instruments, providing for data storage, training data recorders, and reviewing the collected data for completeness and anomalies. The time and expense required for data collection may not be feasible for a modeling project, especially for initial models of a system. In these cases, elicitation of parameter values from domain experts should be considered.

Occasionally domain experts can offer constants or functions for parameter values, but usually recalling values from experience will produce estimates with variation. When consulting experts in a system or process, they can be asked for a most likely value of a parameter, a minimum value, and a maximum value. The definition of minimum and maximum can lead to debates as to whether these mean extreme values (the lowest and highest values either seen or possible) or "90% values." One way to facilitate this discussion is to ask for low and high values as usually occurring and then ask whether extremes beyond the usual lows and highs can appear. The following series of questions is helpful for eliciting data for each parameter.

What is the lowest value of the parameter most of the time?
What is the highest value of the parameter most of the time?
Is a value between these two more likely than either one? If so, what is it?
If the parameter has a most likely value, can values occur that are less than the lowest value or more than the highest value?

Once data is obtained for model inputs, it must be analyzed to produce input values. Input analysis can involve a number of statistical tools, and it is the subject of Chapter 5. It will elaborate more on input analysis methods, including using the answers to the preceding questions. It is sufficient to say that input analysis is another step involved in model iteration. Much effort can be spent choosing input values including input distributions. The best approach in many cases is to apply just enough effort to obtain input values that are good enough to test a model with runs for a sensitivity analysis. The sensitivity analysis varies the input values systematically so that the relative influence of inputs can be estimated. With these estimates, a modeler can decide which inputs deserve further effort for refinement.

2.5 MODEL ASSESSMENT

When a model has been produced, stakeholders call upon the modeler(s) to explain the model's credibility. The modeler must establish his/her own confidence in the model and then convey that confidence to stakeholders and peers, usually through sharing results of Verification and Validation (V&V) exercises. Verification exercises determine whether or not a model is built correctly (error-free) and represents the intended behavior according to the model specification. Validation exercises determine whether the model provides an adequate representation of the real system for the model's stated purpose and addresses the sponsor's problem. The importance of model assessment cannot be understated: stakeholders must have confidence in a model in order to use its results well.

Most textbooks on M&S cover model assessment. Richardson and Pugh, in their text on system dynamics modeling [25], present a model assessment scheme summarized in Table 2.2 that distinguishes exercises for structural assessment and behavioral assessment. They also add model evaluation exercises to the V&V exercises in their scheme outlined below. This assessment scheme is also covered in more detail in modern treatments of system dynamics including [29] and [18].

The details of these assessments are listed below:

Verification of structure

Equation review is a modeler check of every equation for correctness. The parameter dimensions of each equation are analyzed for potential errors. Each equation is checked for the effects of extreme values, including division by zero.

Structural adequacy review is a modeler review to ensure that the elements included in the model and the level of abstraction are sufficient to address the model's stated purpose and specification.

Verification of behavior

Traces of specific entities are essential to verifying a model's logic and the correctness of its implementation. Tracing should cover each path and test execution of each condition.

TABLE 2.2
A Model Assessment Scheme (adapted from Richardson and Pugh [25])

	Structure	Behavior
Verification	Equation review Structural adequacy review	Parameter variability testing Structural insensitivity review Traces
Validation	Face validity review Parameter validity review	Output comparisons Outputs discrimination check Case replication Case prediction
Evaluation	Model appropriateness review	Sensitivity analysis Unexpected behavior review Anomalous behavior review Extreme inputs review Process insights review System improvement test

Parameter variability testing is testing that model outputs reflect changes to model inputs in the same way that the real system outputs reflect changes to real system inputs.

Structural insensitivity review is an expert review of model outputs and behavior against its structure to ensure that the level of abstraction does not inhibit the modeling purpose.

Validation of structure

Face validity is assessed by expert review of the model structure to confirm that, for the modeling purpose, it is an adequate representation of the real system.

Parameter validity is assessed by expert review of parameters to ensure that they are recognizable and not contrived, and that parameter values represent the best available information about the real system.

Validation of behavior

Comparisons of model outputs and system outputs are made statistically. System output values should fall within confidence intervals on the mean model outputs.

Experts comparing system outputs and model outputs should not be able to discriminate between the two.

A case replication demonstrates the model's ability to imitate real system results.

A case prediction demonstrates the model's ability to forecast system behavior accurately. This may be a statistical test in which a prediction interval is constructed and system output is tracked to see whether it falls within the interval.

Evaluation of structure

Members of the intended audience review the model for appropriate levels of abstraction/detail and simplicity/complexity.

Evaluation of behavior

Sensitivity analyses can be run using designed experiments to identify the relative influence of input variables on each outcome, as well as interactions between inputs. The results should be useful in explaining some phenomena in the real system.

A modeler reviews any unexpected behavior found during model usage and performs traces to explain the behavior. If possible, the behavior should be observed in the real system.

If the model exhibits anomalous behavior, it should be reviewed against the real system to ensure that it does not conflict with system behavior.

When extreme values are input to the model, model behavior should imitate system behavior when the system is subjected to the same inputs.

Review model usage and record any insights gained about system behavior.

A system improvement test considers whether the model identifies any system improvements.

Most, but not all, of the foregoing exercises are necessary for building confidence in a model. The verification exercises are essential for the modeler's confidence that a model works correctly. The tracing is especially important so that a modeler can explain the inner workings when called upon to do so. Fortunately, verification exercises can be run iteratively during model construction so that they are not all left until afterward (when stakeholders are calling for modeling results they can use).

Validation and evaluation of structure starts with the first time a modeler presents a draft model to stakeholders, explains it to them, and elicits their feedback. This feedback usually helps correct and refine model structure as well as facilitate further data collection. In such a meeting, the model specification can be refined, data collection can continue, and model structure can be validated.

Validation and evaluation of behavior exercises provide a number of opportunities for demonstrating what a model can do. The exercise that stakeholders request most often is replication of a case of actual results. Therefore, early in the modeling process, a modeler should be looking for a case of recent system behavior that the model can be calibrated to imitate. In the course of such a calibration, if the modeler cannot reproduce one or two primary system outcomes, then the modeler must reexamine data and assumptions, sometimes returning to system experts. For example, it can happen that elicited values for a parameter were inaccurate and that trying new values allows the model to reproduce system outputs. If the experts agree that the initial values may have been erroneous and that new values are better, then the modeler has leeway for calibration.

The foregoing example also illustrates occasions in which a model is like a puzzle: the modeler obtains all the pieces and puts them together. If they do not fit (i.e., outcomes do not imitate real system behavior), then the modeler must figure out which piece(s) do not fit well and scrutinize them with experts. They may find that they tried to insert the wrong piece into the puzzle.

2.5.1 EXPERIMENTATION AND OUTPUT ANALYSIS

As a model is assessed, experimentation and output analysis can begin in earnest. Good evaluation exercises usually signal the start of experimentation and provide the model with insights for further experimentation. Unfortunately, too many models are not used often enough for experimenting, sometimes because the modeler lacks the imagination to create experiments or statistical knowledge and skills to conduct experiments. Modelers are encouraged to put as much effort into using a model as they put into making it.

One school of thought on simulation is that M&S is a learning tool for investigating how systems work. They measure success from what one can learn through M&S. Another school of thought views M&S primarily as a statistical tool that produces quantitative data to support decision making. The two schools of thought are not mutually exclusive, as indicated in the discussion on model specification. Models can have a range of utility and a particular model can evolve from a simple learning model to a decision support model.

Like input analysis, output analysis usually involves statistical analysis of output data and presentation of the data in formats accessible by the stakeholder audience. Analyzed outputs should be used to tell a story and provide information that addresses the problems about which stakeholders are thinking.

2.6 MAKING RECOMMENDATIONS AND REPORTING

As a modeler exercises a model, generates output data, and analyzes it, model behaviors are revealed and he/she comes to understand how a model works. This information provides a basis for discussions with system participants who can apply their knowledge of system workings. System participants often know about the behavior a model describes but the modeler can clarify and quantify the behavior. When they find behavior that is counterintuitive or issues that are unexplained, the modeler is provided opportunities for investigation by running model scenarios.

Recommendations can also emerge from these conversations. Participants often have ideas that they think should be implemented for improving the way a system or process works, and the modeling results may indicate which ideas are most useful. In exploring alternative scenarios, the modeler may also see improvement opportunities that participants had not seen.

Insights and recommendations should not come as surprises to stakeholders. Those surprised by recommendations may be reluctant to accept them. Ongoing discussions should lead up to the final report and briefing with its recommendations.

Good documentation of an M&S study is important for facilitating understanding of the results, especially among those not directly involved in the study, and promoting their implementation. Though a model may not be used by stakeholders, making it accessible confirms transparency of the study. an M&S study report should include a clear description of the problem and the appropriate view of the system, the objectives of the study, modeling specification and assumptions, the modeling process, an overview of the model, the model assessment activities and results, and conclusions with recommendations. The following outline may be used as a checklist for an M&S report.

1. Introduction
    - Problem statement and description
        - Extent and history of the problem, causes of the problem, possible solutions and their obstacles
    - Purpose of study
    - Purpose of model building
    - Executive summary of key results
2. Background
    - System description
        - Diagrams that illustrate the system configuration
    - System reference behavior
        - Narrative description, tabular data, graphs
    - Assumptions and the underlying rationale for each
3. Model Development
    - Modeling process
        - Modeling approach and sequence of events
    - Model evolution
    - Data acquisition
        - Source of data, method of collection, problems, solutions, analysis, etc.
4. Model Description
    - Time frame, spatial boundaries, entities, attributes, key assumptions
    - Process flow and model structure
        - Flowcharts, model diagrams, etc.
    - Key logic and equations, sources of uncertainty, probability distributions, etc.
    - Model Overview
        - Sources of data and input analysis
        - Flowcharts, flow networks, block diagrams, etc.
    - Assumptions and other model details
    - Approach for verification and validation
5. Model Verification and Validation
    - Testing results, extreme values, sensitivity analysis
    - Comparison to reference behavior, expert review, etc.
    - Statistical analyses

Other testing and V&V methods

6. Model Application and Transition
    Experimentation and analysis of results
    Tabulated data, plots, bar graphs, histograms, pie charts, statistical analyses and other reports, etc.
    Interpretation of model results
    Limitations of the model and future enhancements
    Next steps
    Model transfer issues
        Availability of model user documentation and of model developer documentation
7. Conclusions and Recommendations
    Real-world system
        Policy suggestions
    Future model usage and development
    Modeling process notes
    Modeling tool(s) and platform used
    Process improvement scenarios
    Future research
8. Appendices (if necessary)
    Supplementary model details
    Model run output
    Selected computer outputs, related letters, technical articles, etc.
    Additional data analysis as needed

2.7 KNOWLEDGE AND SKILLS FOR MODELING AND SIMULATION

The discussion in this chapter should have made clear that M&S requires knowledge and skills, so this last section discusses these areas explicitly. A modeler must possess knowledge and skills in four areas. First, one must understand modeling and simulation concepts, processes, paradigms, and tools as described in this chapter. M&S tools are numerous and are evolving, and full descriptions are beyond the scope of this book. Reviews and surveys of simulation software are readily available online. An example survey of tools kept updated is at `http://www.orms-today.org/surveys/Simulation/Simulation.html`.

The second area in which a modeler requires knowledge is that of the domain being modeled. The domain might be distinguished in two degrees of relevance: specific systems and fields. A modeler who is familiar with specific systems may also be a domain expert capable of modeling with minimal reliance on other system experts. A modeler who is expert in a field, for example product development, and is asked to model a specific development program must rely heavily on experts in the development program to acquaint him/her with its specific structure and data. In this case, a modeler brings familiarity with the structure and dynamics of product development as well as expertise in representing them. Sometimes a modeler is requested for his/her M&S skills despite a lack of familiarity with a field. In this case, he/she may learn from published modeling work in the field, but is still more reliant on domain experts to learn about the field and the subject system or process in sufficient depth for modeling.

Engagement skills are a third area in which a modeler should be well-versed. This area includes human relations, organizational, and project management skills that enable a modeler to meet with potential customers; assess their needs and relate them to possible M&S solutions; develop reasonable expectations for an engagement; negotiate a program of study; work with participants to specify a model, collect data, and validate the model; keep sponsors informed; and report useful results.

The fourth M&S knowledge and skill area for modelers is statistics. Models are essentially quantitative tools executed on computers and the success of most M&S studies depends on quantitative analysis. Input values are usually the result of statistical analysis, such as distribution fitting. Model experimentation and sensitivity analysis often require designed experiments that are analyzed with analysis of variance (ANOVA). Outputs may also need to be analyzed with hypothesis testing, regression analysis, confidence intervals, and prediction intervals. The modeler uses statistics to understand system behavior as represented in a model and relate a story of that behavior to stakeholders. See statistics details in Chapters 4, 5, and 6.

In summary, keys to a successful M&S project include a well-defined and achievable goal, complementary skills on the project team, an adequate level of user participation, selection of an appropriate simulation toolset or language, and effective project management.

2.8 SUMMARY

Modeling and simulation (M&S) projects vary widely, so modeling processes and methodologies can be employed very differently across projects. The nominal modeling stages are engagement, specification, construction, verification and validation, experimentation, and reporting. But the modeling process does not proceed through these stages in a strictly sequential manner because a modeler undertakes combinations of these stages iteratively. A modeler needs specialized knowledge and skills to conduct a modeling process well.

First, one should identify the stakeholders including sponsors, process or system owners, and the process or system participants who do the actual work. Stakeholders may also include other engineers and scientists in related disciplines using the simulation results. The modeler must always keep stakeholders informed of the study and its progress.

Model specification begins with identifying a problem and describing its context for a modeling purpose. The purpose will include the intended audience for the modeling results, the desired set of policies to be simulated, and the kinds of decisions the results are intended to support. The modeler must choose an appropriate level of model abstraction necessary to answer the questions. By default start with a simple model and add detail as necessary.

One of the first decisions in modeling is choosing the modeling paradigm. The different paradigms of continuous modeling, discrete event simulation, and agent-based simulation have respective advantages and limitations for various types of modeling.

Continuous models are usually more abstract than discrete event models. Use continuous modeling if a global view is desired and aggregated entities are sufficient. Discrete event models can be easier to understand for audiences that think of entities being changed through activities performed on them, and useful for investigating system-level process behavior with visibility into individual entities. Agent-based modeling is also useful for disaggregated modeling whereby individual object behavior can be described and if one is interested in collective behavior that emerges from their interactions. The choice of modeling paradigm may also depend on required simulation capabilities and the modeler's expertise.

During model construction, decisions must be made about representing the real-world subject. For continuous models, one must decide what flows to represent, how they are modified, and how they interact. For discrete event models, one must decide how to represent system elements using entities and resources, and how entities are routed. For agent-based models, agents are identified, their characteristics are specified, and rules for their behavior are written.

Another structural decision in model construction is the partitioning of a model. An iterative approach with successive elaborations ("build a little, test a little") is best to support these modeling decisions. Iterative construction also facilitates assessment of model performance as it is being built, and configuration management of the ongoing model changes.

For data collection and input analysis, the input parameters should be created such that the data can be collected and outputs defined that will be available for validation. The costs of data collection should be considered, and elicitation of parameter values from domain experts may be warranted.

Model assessment addresses 1) verification to determine whether or not a model is built correctly without errors and represents the intended behavior according to the model specification, and 2) validation to determine whether the model provides an adequate representation of the real system for the model's stated purpose and addresses the sponsor's problem. V&V covers both structural and behavioral aspects.

Experimentation and output analysis can begin when a model is being assessed. Like input analysis, output analysis involves statistical analysis of output data and presentation of it to the stakeholder audience such that it addresses their problems.

Good documentation of an M&S study is important for facilitating understanding of the results, especially among those not directly involved in the study, and promoting their implementation.

A modeler must possess the right knowledge and skills to conduct a good simulation study. One must understand M&S concepts, processes, paradigms, and tools; have knowledge of the application domain; engagement skills to work with people; and statistics skills for the quantitative analysis.

The overall keys to a successful M&S project include a well-defined and achievable goal, complementary skills on the project team, an adequate level of user participation, selection of an appropriate simulation toolset or language, and effective project management.

3 Types of Simulation

This chapter presents an overview of the major types of simulation models with simple illustrations. The systems and/or their models can be discrete, continuous, or a combination. The essential difference in modeling is how the simulation time is advanced. Continuous systems modeling methods such as system dynamics always advance time with a constant delta (dt), a fixed time slice. Since variable values may change within any time interval in a continuous system, the delta increment is very small and time-dependent variables are recomputed at the end of each time increment. The variable values change continuously with respect to time. In contrast, in discrete modeling, the changing of variable values is normally event-based. State changes occur in discrete systems at aperiodic times depending on the event happenings. The simulation time is advanced from one event to the next in a discrete manner.

All classes of systems may be represented by any of the model types. A discrete model is not always used to represent a discrete system and vice versa. The choice of model depends on the specific objectives of a study.

Simulation software may be oriented for specific model types or allow hybrid approaches. In general, the software tools must implement the following functions:

- Initialization of system variables
- Time clock for flow control
- For continuous systems:
  - Time advancement in equal small increments, with all variables recomputed at each time step
- For discrete systems:
  - Event calendar for timing
  - Time advancement from one event to the next, with no changes in variables between events
  - Event processing routines
- Statistics collection
- Output generation

Additional features beyond the above may include a graphical user interface, animation, a variety of probability distributions for random variates, design of experiments, optimization, interfaces with other software or databases, and more. See Appendix A for some available simulation tools to support continuous, discrete event, agent-based, and hybrid modeling applications.

Models may be deterministic (having no probabilistic variables) or stochastic (including probabilistic variables). Few engineering applications are wholly deterministic. When a model has probabilistic variables, each run can have a different outcome constituting an estimate of the model characteristics. Therefore, many runs must be made to characterize output variables. See Chapter 4 on handling randomness in models, Chapter 5 for stochastic inputs, and Chapter 6 for output analysis of stochastic systems to address uncertainty.

3.1 CONTINUOUS

Continuous systems models consist of differential or algebraic equations to solve variables representing the state of the system over time. A continuous model shows how values change as functions of time computed at equidistant small time steps. These changes are typically represented by smooth continuous curves. Continuous simulation models may also be applied to systems that are discrete in real life but where reasonably accurate solutions can be obtained by averaging values of the model variables.

Continuous models perform a numerical integration of differential equations over time. A practical and simple solution is to use Euler's method. It approximates the derivative of the function $y(t)$ by using the forward difference method. It computes the following at each time increment:

$$y(t_{n+1}) = y(t_n) + \frac{dy}{dt}(t_n)\Delta t \tag{3.1}$$

where

y is a function over time
$\frac{dy}{dt}$ is the time derivative of y
n is the time index
Δt is the time step size to increment t_n to t_{n+1}

The Runge-Kutta method uses additional gradients within the time divisions for more accuracy. See Appendix A for simple programs for continuous systems demonstrating Euler's method and the Runge-Kutta method. These programs use integration methods for computing variable values at each time step within a time loop.

System dynamics is the most widely used form of continuous simulation. System dynamics refers to the simulation methodology pioneered by Jay Forrester, which was developed to model complex continuous systems for improving management policies and organizational structures [5] [6]. Improvement comes from model-based understandings.

System dynamics provides a very rich modeling environment. It can incorporate many formulations including equations, graphs, tabular data, or otherwise. Models are formulated using continuous quantities interconnected in loops of information feedback and circular causality. The quantities are expressed as levels (also stocks or accumulations), rates (also called flows), and information links representing the feedback loops.

Levels represent real-world accumulations and serve as the state variables describing a system at any point in time (e.g. the number of cars on a road, number of manufacturing defects, height of water in a dam, work completed, etc.). Rates are the flows over time that affect the levels. See Table 3.1 for a description of model elements.

The system dynamics approach involves the following concepts [24]:

- Defining problems dynamically, in terms of graphs over time
- Striving for an endogenous (caused within) behavioral view of the significant dynamics of a system
- Thinking of all real systems concepts as continuous quantities interconnected in information feedback loops and circular causality
- Identifying independent levels in the system and their inflow and outflow rates
- Formulating a model capable of reproducing the dynamic problem of concern by itself
- Deriving understandings and applicable policy insights from the resulting model
- Ultimately implementing changes resulting from model-based understandings and insights, which was Forrester's overall goal

A major principle of system dynamics modeling is that the dynamic behavior of a system is largely a consequence of its structure. Thus, system behavior can be changed not only by changing input values but by changing the structure of a system. Improvement of a process thus entails an understanding and modification of its structure. The structures of the as-is and to-be processes are represented in models.

The existence of process feedback is another underlying principle. Elements of a system dynamics model can interact through feedback loops, where a change in one variable affects other variables over time, which in turn affects the original variable. Understanding and taking advantage of feedback effects in real systems can provide high leverage.

3.1.1 CONSERVED FLOWS AND THE CONTINUOUS VIEW

In the system dynamics worldview, individual entities are not represented as such. Instead they are abstracted into aggregated flows. The units of flow rate are the number of entities flowing per unit of time. These flows are considered material or physical flows that must be conserved within a flow chain.

Information links connect data from auxiliaries or levels to rates or other auxiliaries. Information connections are not conserved flows because nothing is lost or gained in the data transfer.

A physical example of a real-world level/rate system is a water network, such as a set of holding tanks connected with valved pipes. It's easy to visualize the rise and fall of water levels in the tanks as inflow and outflow rates are varied. The amount of water in the system is conserved within all the reservoirs.

The continuous view does not track individual events, rather flowing entities are treated in the aggregate and systems can be described through differential equations.

There is a sort of blurring effect on discrete events. The focus of continuous models is not on specific individuals or events like discrete event approaches (described in Section 3.2), but instead on the patterns of behavior and on modeling average individuals in a population.

3.1.2 MODEL ELEMENTS

System dynamics model elements are summarized in Table 3.1. Their standard graphical notations are shown in the next example model. The same notations with similar icons are used consistently in the modeling tools for system dynamics listed in Appendix A.

TABLE 3.1: System Dynamics Model Elements

Element	Description
Level	A level is an accumulation over time, also called a stock or state variable. It can serve as a storage device for material, energy, or information. Contents move through levels via inflow and outflow rates. Levels represent the state variables in a system and are a function of past accumulation of rates.
Source/Sink	Sources and sinks indicate that flows come from or go to somewhere external to the system or process. Their presence signifies that real-world accumulations occur outside the boundary of the modeled system. They represent infinite supplies or repositories that are not specified in the model.
Rate	Rates are also called flows; the actions in a system. They effect the changes in levels. Rates may represent decisions or policy statements. Rates are computed as a function of levels, constants, and auxiliaries.
Auxiliary	Auxiliaries are converters of input to output, and help elaborate the detail of stock and flow structures. An auxiliary variable must lie in an information link that connects a level to a rate. Auxiliaries often represent score-keeping variables.

(*continued*)

TABLE 3.1: System Dynamics Model Elements (*continued*)

Element	Description
Information Link	Information linkages are used to represent information flow (as opposed to material flow). Rates, as control mechanisms, often require connectors from other variables (usually levels or auxiliaries) for decision making. Links can represent closed-path feedback loops between elements.

Elements can be combined together to form larger infrastructures with associated behaviors. Using common existing infrastructures in models can save a lot of time and headache when reusing them. An infrastructure can be easily modified or enhanced for different modeling purposes. See [18] for a taxonomy of infrastructures with examples that apply to engineering contexts.

3.1.3 MATHEMATICAL FORMULATION OF SYSTEM DYNAMICS

This section explains the mathematics of system dynamics modeling for general background, and is not necessary to develop or use system dynamics models. It illustrates the underpinnings of the modeling approach and provides a mathematical framework for it. Knowing it can be helpful in constructing models. An elegant aspect of system dynamics tools is that systems can be described visually to a large degree, and there is no need for the user to explicitly write or compute differential equations. The tools do all numerical integration calculations. Users do, however, need to compose equations for rates and auxiliaries, which can sometimes be described through visual graph relationships describing two-dimensional (x-y) plots.

The mathematical structure of a system dynamics simulation model is a set of coupled, nonlinear, first-order differential equations per Equation 3.2.

$$x'(t) = f(x, p) \tag{3.2}$$

where

x is a vector of levels

p is a set of parameters

f is a nonlinear vector-valued function

State variables are represented by the levels. As simulation time advances, all rates are evaluated and integrated to compute the current levels. Runge-Kutta or Euler's numerical integration methods are normally used. These algorithms are described in standard references on numerical analysis methods or provided in technical documentation with system dynamics modeling tools. Also see the software examples in Appendix B.

Numerical integration in system dynamics determines levels at any time t based on their inflow and outflow rates per Equation 3.3, where the dt parameter is the chosen time increment.

$$Level(time) = Level(time - dt) + (inflow - outflow) * dt \tag{3.3}$$

Describing the system with high-level equations spares the modeler from integration mechanics. Note that almost all tools also relieve the modeler of constructing the equations; rather, a diagrammatic representation is drawn and the underlying differential equations are automatically produced.

Example: System Dynamics Model for Resource Allocation

Engineering management needs to allocate personnel resources based on relative needs among competing tasks. The model in Figure 3.1 contains resource allocation infrastructures to support this strategic decision making. It is implemented with the iThink tool (see Appendix A). The resource allocation model has levels, shown as boxes, for the tasks (in work units) and task resources (number of people). The two primary flows represent two streams of work: each set of work enters a backlog, is performed at a rate illustrated by the valves, and then enters a completed state. Auxiliary variables are designated as circles (e.g. *Task 1 Productivity*), sources and sinks are the clouds, and information links are the single arrows connecting elements.

In each task flow chain, the rate of work accomplished is represented as a generic production process shown in Figure 3.2. This is an example of a well-established infrastructure. The flow rate drains the task backlog level and accumulates into completed work. The production rate of work completion is computed by multiplying the resources (# of people) by their average productivity (tasks per person per time unit).

Tasks with the greatest backlog of work receive proportionally greater resources ("the squeaky wheel gets the grease"). The allocation infrastructure will adjust dynamically as work gets accomplished, backlogs change, and productivity varies. The dynamic behavior is driven by the equations for the continuously adjusted proportion of resources allocated for a task per Figure 3.3. Resources are allocated dynamically based on each workflow's backlog. More resources are allocated to the workflow of tasks having the larger backlog. As work is accomplished, backlogs change and productivity varies.

Productivity serves as a weighting factor to account for differences in the work rates in calculating the backlog effort. Auxiliaries are used to help compute the estimated effort for each task by adjusting for productivity. The full model equations are shown in Figure 3.4. They include the standard graphic icons for system dynamic elements of rates, levels, and auxiliaries. Graphic curves show defined time-varying input functions (also listed with their table values).

This type of resource policy is used frequently when creating project teams, allocating staff, and planning their workloads. With a fixed staff size, proportionally more people are dedicated to the larger tasks. For example, if one task is about twice as large as another, then that team will have roughly double the people. The work

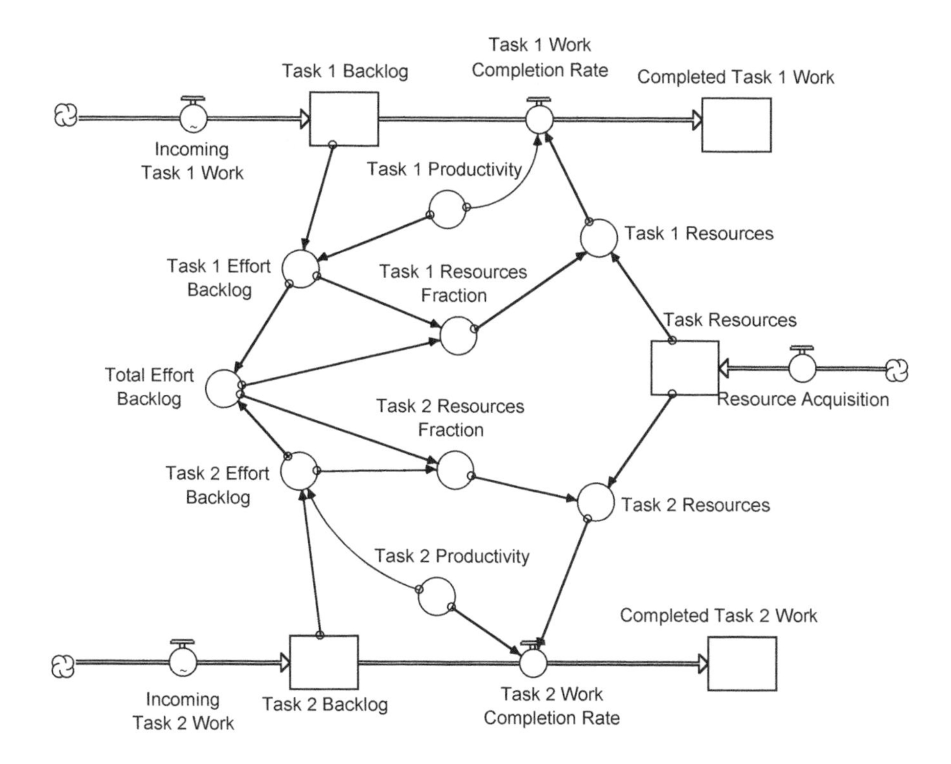

FIGURE 3.1: Resource Allocation Model

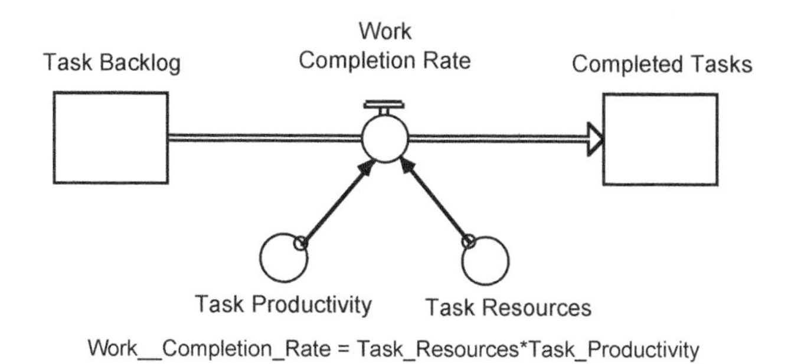

FIGURE 3.2: Task Production Structure and Rate Equation

backlogs are monitored throughout a project, and people re-allocated as the work ratios change based on the respective estimates to complete. This example models the allocation between two tasks, but it can be easily extended to cover more competing tasks. Many more variations of the resource allocation infrastructure are also possible.

Figure 3.5 shows the results of a simulation run where both tasks start with equal

```
Task_1_Effort__Backlog = Task_1_Backlog/Task_1_Productivity
Task_2_Effort__Backlog = Task_2_Backlog/Task_2_Productivity
Total_Effort__Backlog = Task_1_Effort__Backlog+Task_2_Effort__Backlog
Task_1_Resources__Fraction = Task_1_Effort__Backlog/Total_Effort__Backlog
```

FIGURE 3.3: Task Resource Allocation Equations

```
□ Completed_Task_1_Work(t) = Completed_Task_1_Work(t - dt) + (Task_1_Work__Completion_Rate) * dt
  INIT Completed_Task_1_Work = 0
  INFLOWS:
    Task_1_Work__Completion_Rate = Task_1_Resources*Task_1_Productivity
□ Completed_Task_2_Work(t) = Completed_Task_2_Work(t - dt) + (Task_2_Work__Completion_Rate) * dt
  INIT Completed_Task_2_Work = 0
  INFLOWS:
    Task_2_Work__Completion_Rate = Task_2_Resources*Task_2_Productivity
□ Task_1_Backlog(t) = Task_1_Backlog(t - dt) + (Incoming__Task_1_Work - Task_1_Work__Completion_Rate)
  * dt
  INIT Task_1_Backlog = 10
  INFLOWS:
    Incoming__Task_1_Work = GRAPH(time)
    (1.00, 1.00), (2.00, 1.00), (3.00, 1.00), (4.00, 1.00), (5.00, 1.00), (6.00, 1.00), (7.00, 1.00), (8.00, 1.00),
    (9.00, 1.00), (10.0, 1.00), (11.0, 1.00), (12.0, 1.00), (13.0, 1.00)
  OUTFLOWS:
    Task_1_Work__Completion_Rate = Task_1_Resources*Task_1_Productivity
□ Task_2_Backlog(t) = Task_2_Backlog(t - dt) + (Incoming__Task_2_Work - Task_2_Work__Completion_Rate)
  * dt
  INIT Task_2_Backlog = 10
  INFLOWS:
    Incoming__Task_2_Work = GRAPH(time)
    (1.00, 1.00), (2.00, 1.00), (3.00, 1.00), (4.00, 1.00), (5.00, 1.00), (6.00, 2.02), (7.00, 1.98), (8.00, 2.00),
    (9.00, 1.00), (10.0, 1.00), (11.0, 1.00), (12.0, 1.00), (13.0, 1.00)
  OUTFLOWS:
    Task_2_Work__Completion_Rate = Task_2_Resources*Task_2_Productivity
□ Task_Resources(t) = Task_Resources(t - dt) + (Resource_Acquisition) * dt
  INIT Task_Resources = 4
  INFLOWS:
    Resource_Acquisition = 0
○ Task_1_Effort__Backlog = Task_1_Backlog/Task_1_Productivity
○ Task_1_Productivity = .5
○ Task_1_Resources = Task_Resources*Task_1_Resources__Fraction
○ Task_1_Resources__Fraction = Task_1_Effort__Backlog/Total_Effort__Backlog
○ Task_2_Effort__Backlog = Task_2_Backlog/Task_2_Productivity
○ Task_2_Productivity = .5
○ Task_2_Resources = Task_Resources*Task_2_Resources__Fraction
○ Task_2_Resources__Fraction = Task_2_Effort__Backlog/Total_Effort__Backlog
○ Total_Effort__Backlog = Task_1_Effort__Backlog+Task_2_Effort__Backlog
```

FIGURE 3.4: Resource Allocation Model

backlogs and 50% allocations due to equal incoming work rates. At $time = 6$ more incoming work starts streaming into the backlog for Task 2 as seen on the graph. The fractional allocations change, and more resources are then allocated to work off the excess backlog in Task 2. The graph for *Task 2 Resources Fraction* increases at that point and slowly tapers down again after some of the Task 2 work is completed.

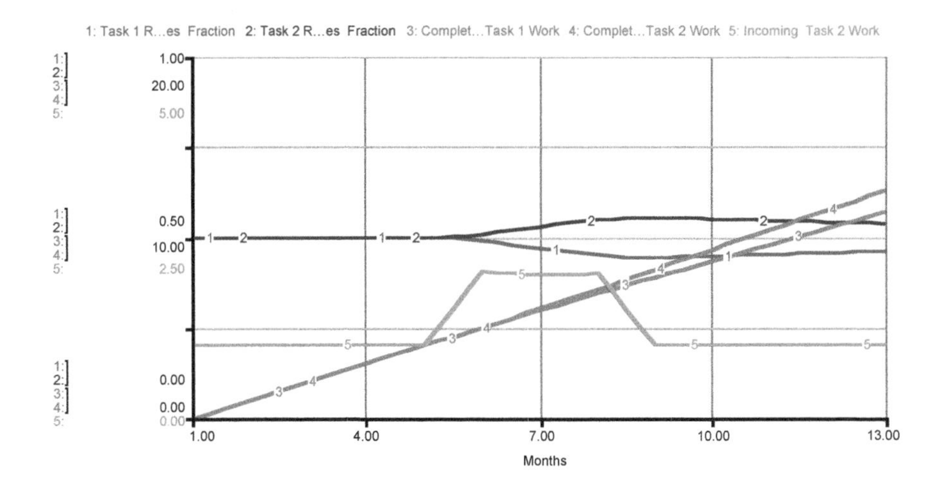

FIGURE 3.5: Resource Allocation Model Output

3.2 DISCRETE EVENT

Discrete event models consist of flows of individual entities, characterized by the values of their respective attributes. The characteristics of the entities can change during a simulation run. These changes occur instantaneously as the simulated time lapses. The entities move through a system represented as a network of nodes, perform activities by using resources, and create events that change the state of a system. Standard discrete event model elements are summarized in Table 3.2. They are described in more detail in subsequent sections.

TABLE 3.2: Discrete Event Model Elements

Element	Description
Create node	Generator of entities to enter a system.
Terminate node	Departure point of entities leaving the system.
Activity node	Locations where entities are served and consume resources.
Entity	An object in a system whose motion may result in an event.
Resource	Commodities used by entities as they traverse in a system.

(continued)

TABLE 3.2: Discrete Event Model Elements (*continued*)

Element	**Description**
Path	Routes that entities travel between nodes.
Batch/Unbatch Nodes	Where entities are combined into groups or un-combined.
Information link	Data connections between model elements for logical and mathematical operations.

3.2.1 NEXT EVENT TIME ADVANCE APPROACH

This section describes the discrete event-based approach for a queue/server system and generalizes the introductory example calculations in Section 1.6.1. It illustrates the logic used for time stepping through an event-based simulation, updating of statistics, etc. The next-event time-advance approach for a single-server queueing system will be described using the following notation and illustrated in Figure 3.6. It is the standard procedure used in discrete event modeling. This description (derived from [17]) will step through the computation sequence of a simulation including the generation of random numbers as normally performed by a simulation tool. The nomenclature used is:

t_i = time of arrival of the ith customer
$A_i = t_i - t_{i-1}$ = interarrival time between $(i-1)$st and ith arrivals of customers
S_i = time that server spends serving ith customer
D_i = delay in queue of ith customer
$c_i = t_i + D_i + S_i$ = time that ith customer completes service and departs
e_i = time of occurrence of ith event

The variables are random numbers and we assume that the probability distributions of the interarrival times A_1, A_2,... and the service times S_1, S_2,... are known. Their cumulative distribution functions can be used to generate the random variates per the methods described in Chapter 4.

The event times, e_i's, are the values the simulation clock takes on after $e_0 = 0$. At $time = 0$ the server is idle. The time t_1 of the first arrival is determined by generating A_1 and adding it to 0. The simulation clock is then advanced from e_0 to the next event time e_1. In Figure 3.6 the curved arrows represent advancing the simulation clock.

The first customer arriving at time t_1 finds the server idle and immediately enters service with a delay in queue of $D_1 = 0$. The status of the server is changed from idle to busy. The time c_1 when the customer will complete service is computed by generating the random variate S_1 and adding it to t_1.

The time of the second arrival, t_2, is computed as $t_2 = t_1 + A_2$, where A_2 is another generated random arrival time. If $t_2 < c_1$ as shown in Figure 3.6, the simulation clock is advanced from e_1 to the time of the next event, $e_2 = t_2$. If c_1 was less than t_2, the clock would be advanced from e_1 to c_1.

The customer arriving at time t_2 finds the server already busy. The number of customers in the queue is increased from 0 to 1 and the time of arrival of this customer is recorded.

Next the time of the third arrival, t_3, is computed as $t_3 = t_2 + A_3$. If $c_1 < t_3$ as depicted in the figure, the simulation clock is advanced from e_2 to the time of the next event, $e_3 = c_1$, where the first customer completes service and departs. The customer in the queue that arrived at time t_2 now begins service. The corresponding delay in queue and service completion times are computed as $D_2 = c_1 - t_2$ and $c_2 = c_1 + S_2$ while the number of customers in the queue is decreased from 1 to 0.

If $t_3 < c_2$ the simulation clock is advanced from e_3 to the next event time $e_4 = t3$ as shown in Figure 3.6. If c_2 was less than t_3, the clock would be advanced from e_2 to c_2, etc.

With the sequence of events per Figure 3.6, it can be seen that the server is idle between times c_2 and t_3 before the third customer begins service.

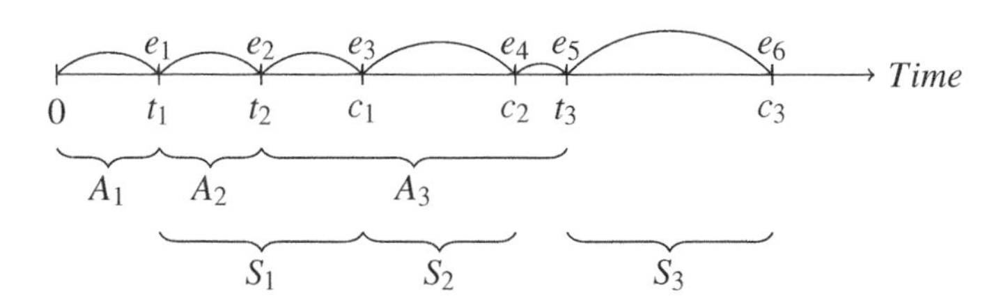

FIGURE 3.6: Next Event Time Advance Approach

3.2.2 COMMON PROCESSES AND OBJECTS IN DISCRETE SYSTEMS

Discrete event simulation views systems and processes as interconnected event-based flows of entities through queues and activities. This view corresponds well with intrinsic, measurable, real-world phenomena. Using this view, discrete event simulation represents system behavior using a set of common constructs, described in this section primarily following the concepts and descriptions in [16]. Capabilities to represent these are to be expected of discrete event modeling tools.

In graphical modeling tools these processes may be represented as visual icons called nodes, blocks, modules, or similar names. They may be termed functions or modules in a modeling language. In all cases they represent a model piece or construct with associated behavior logic.

The moving objects in a system whose motion results in events are typically called entities or items. Interconnected nodes represent entity movement routes. The broad similarity of tools is based on movement and accumulation of entities along these connecting paths through the nodes. The modeler logically connects the nodes to represent the entity flows in a system. Model construction entails selection and linking of appropriate model pieces and specifying their parameters.

Knowledge of common processes helps the modeler understand system behavior as well as expected features of modeling tools. The modeler can better identify desirable modeling capabilities for specific problems. Later in this section are graphical and programmed examples of these capabilities with representative and widely used tools in the field. Additional references for more details on discrete event simulation can be found in [17], [2], and [16].

Entity Movement

Dynamism in discrete systems is caused by the movement of entities which results in the occurrence of events that change the system state over time. In most systems the entities enter the system as inputs through the system boundary, move between the components within the system, and may leave the system boundary in the form of output from the system. In other cases the system may initially contain some entities that move through the system and possibly leave. Some entities may never leave the system boundary.

Entity Creation

To create entities entering a system (e.g. cars arriving for service or manufacturing pieces arriving at a facility) a process for the entity source is required. The simulation module or node type representing this process may be called a *source*, *create*, *generate*, etc. A creation node generates entities with desired timings. It should allow for setting the time of creation of the first entity and the times between entity arrivals as either constant or random variables. For the latter a random variate generator provides samples from probability distribution functions by specifying values of its parameters.

Entities may be assigned names when they are generated. This attribute may be used for entity routing through branches on the basis of the entity type, or for specialized treatment of different entities.

Starting and stopping criteria for creation of entities can be dictated. The creation module should allow for stopping the simulation either after a certain amount of time has elapsed or after a specified number of entities. It is also desirable to allow for the batch creation of entities, i.e., the creation of more than one entity at each arrival time.

One may need to initialize the system with entities already existing at accumulation points. Nodes or modules that correspond to the accumulation points (queues) should provide for initialization of the desired number of entities in each queue.

Entity Termination

Entities that enter a system may also leave. A provision for entity departure may be called a *terminate*, *sink*, *depart*, *out*, or similar module name in modeling tools to represent the path end for an entity. A simulation may end with some, all, or none of the entities remaining in the system depending on creation and termination parameters. Entity termination may be a simulation stopping criteria. One may specify the number of entity terminations required to end the simulation run.

Entity Traversal

Entities traverse through paths connecting system components. The paths may have associated delay times or their capacities may be limited. An explicit delay object or node may used to generate a delay corresponding to the travel time of an entity from one node to the other. The delay times may be specified as a constant, a random variable, a user variable, another expression, or be a function of an entity attribute. Path routes can be selectable based on logical conditions, probabilities, or the status of alternative queues on different paths.

Entities traversing a system may also undergo multiplication into more entities, or several entities join to or from a single entity. Carriers can be used to transport batched entities through the network. Grouped items may traverse a carrier entity (e.g. public transit vehicle) and subsequently ungroup. Previously batched entities may become ungrouped, e.g. the unloading of packages from a carrier. Entities that are grouped or assembled together may lose their individual attributes. There may be grouping conditions based on the number of entities or their attributes. An example of different entity types joining would be the assembling of manufactured items from smaller heterogeneous pieces.

Entity Use of Resources

Resources are commodities used by entities as they traverse a system. They represent servers, operators, machines, etc. and have associated service times. They provide services where entities remain while being processed. Arriving entities must wait until the server is free. Resources (sometimes called facilities) reside at fixed system locations and are made available to entities as needed. Resources usually have a base stock with a fixed level of resource units (e.g. serving capacity).

At event times the entities seize resources, use them with an associated delay, and the resources generally become available again for the next entity. Entity priority schemes are possible, such as special entities that jump to the head of a queue or interrupt another in service. There may also be parallel servers available for entities, each possibly with different service characteristics.

Entity Accumulation (Queues)

Entities move through the system and may wait at accumulation points. In discrete systems these accumulation points are called *queues* as the entities line up in order.

Queues form when entities need resource facilities that may be busy or otherwise unavailable. A queue represents a buffer before a resource or facility.

Priority schemes may be specified for leaving the queue to be served (e.g. First-In-First-Out (FIFO), Last-In-First-Out (LIFO) and others). Multiple queues may be formed for parallel servers. There may be a queue waiting-area capacity limit constraint, or threshold waiting times may be specified for entities to depart a queue.

Entities may also accumulate and form queues at logical gates or switches on paths where entities require permission to proceed. Rules may be specified before entities enter service, such as a required matching with another entity type. An example is an assembly operation for incoming entities that are synchronized and integrated to form new entities.

Auxiliary Operations

Auxiliary operations used in discrete event modeling include handling and utilization of variables, manipulation of entities (e.g. transfer, delete, copy), file operations and others. Custom user variables may be specified along with default system variables.

Statistics specification provides for measuring and assessing system performance data. Statistics include both observation-based (e.g. waiting times) and time-based statistics (e.g. utilization) as covered in Sections 1.6.1 and 6.5.1. Capturing of entity event times for arrival, departure, etc., allow for measures such as waiting time or queue length. Server-based statistics include resource utilization measures. For all of these it may be possible to specify only transient or steady-state data collection.

Visual tracing of entities along routes and other animation are additional useful capabilities. These may be helpful to the modeler in debugging and model verification, or external validation and presentation with stakeholders.

Example: Discrete Processes and Model Objects

Figure 3.7 shows a graphical ExtendSim [13] model of a single server queue demonstrating basic capabilities for modeling discrete processes with model blocks. Entities are created, they traverse the system, accumulate, use a resource, and terminate. An operation for generating random numbers for event timing is also included.

Entities are generated in the *Create* block and transition to the *Queue* block awaiting service in the *Activity* block. The service time is dictated by a random number sampled from the *Random Number* block and the entities leave the system through the *Exit* block. See further annotations in the figure. Also see Chapter 7 for an applied case study using ExtendSim for modeling discrete processes.

Example: Car Charging Station Model with SimPy

Discrete event modeling and simulation will be illustrated in detail for the electric car charging station scenarios. This example is elaborated here and in subsequent chapters to demonstrate basic discrete event model components, key statistics and illustrate the overall modeling process steps. The examples will implement modeling

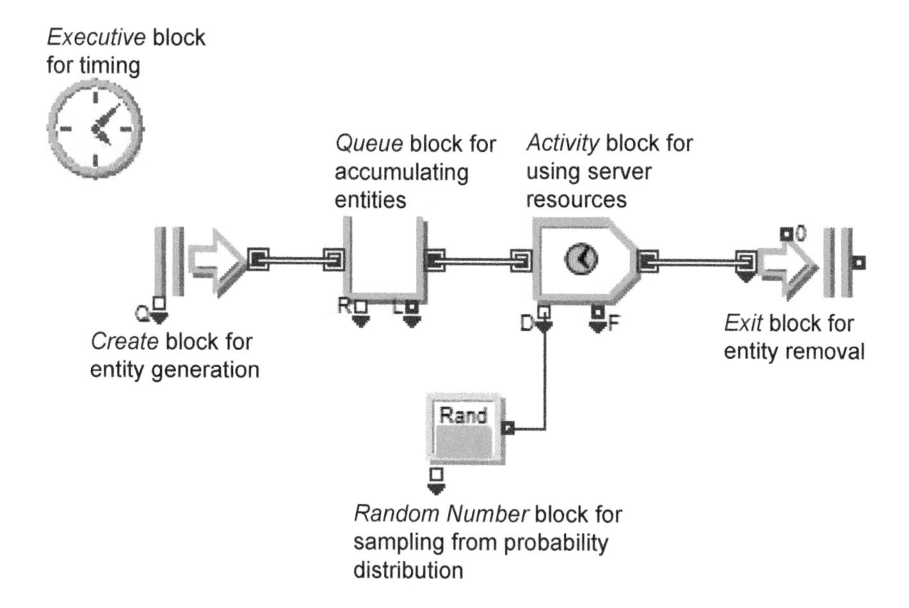

FIGURE 3.7: ExtendSim Model for Single Server Queue with Common Process Blocks

constructs using the SimPy discrete event simulation library written in the Python language. It is open source and accessible to readers to reproduce these examples, extend the source code provided in this chapter and Appendix B, and use for other applications. See [27] for additional details and downloading of SimPy. An independent technical overview of discrete event simulation using SimPy can also be found in [20].

SimPy is a process-oriented discrete event simulation library, so the behavior of active entities like cars or customers is modeled with *processes*. All processes live in an *environment*. They interact with the environment and with each other via *events*.

Processes are described by simple generators that declare functions that behave like an iterator, i.e., they can be used in a loop. During their lifetime, the processes create events and *yield* them waiting to be triggered. When a process yields an event, the process gets suspended. SimPy resumes the process when the event occurs (gets triggered).

A model for the charging station will require instantiating the SimPy simulation environment, defining an electric car process (generator), and a resource for the charging station. We will illustrate building up the model from the beginning and incrementally add complexity. In the subsequent code examples, any text following the "#" character are embedded comments.

Initializing the SimPy simulation environment can be done as in Figure 3.8. The first statement in the source code imports the SimPy library in order to use its functionality. Creating an instance of the `simpy.Environment` will be needed for a car process and thus will be passed into the car process function. The resources used by car entities will also be attached to it.

```
import simpy

# instantiate execution environment for simulation
environment = simpy.Environment()
```

FIGURE 3.8: Initializing Simulation Environment

The execution environment is a class object with associated methods that will be used. The methods use a dot notation, e.g. `environment.now` refers to the current simulation time.

The electric car objects will first need a resource for the charging station. The resource can be defined per Figure 3.9 with the `simpy.Resource` method specifying the resource environment and characteristics. The resource is parameterized by its capacity (e.g. `capacity=1` refers to a single charging bay). This capacity will be varied later for simulation experiments with different operating scenarios.

```
# charging station resource
charging_station = simpy.Resource(environment, capacity
    =1)
```

FIGURE 3.9: Defining Charging Station Resource

The electric car (event) behavior must also be defined before creating actual car instances. The cars will arrive, use the charging resource, and leave. The car process generator is listed in Figure 3.10 defining the events and is attached to the previously defined charging station resource.

This function in Figure 3.10 generates cars during the simulation with their instrinic process behaviors described in the code for arriving, charging, and departing events. Other actions for statistics collection or output statements can also go in the generator. The car process generator also requires a reference to the named `environment` in order to create new events in it.

The first `yield` statement suspends execution until the arrival time occurs, and then triggers the event for the charging station resource. The resource is requested via the `yield request` statement, and the car will enter service when the charging resource is available. If the server is free at arrival time, then it can start service immediately and the code moves on to the next `yield` for the charging time end event. If the server is busy, the car is automatically queued by the charging resource.

```
# electric car process generator
def electric_car(environment, name, charging_station,
    arrival_time, charging_time):
      # trigger arrival event at charging station
      yield environment.timeout(arrival_time)

      # request charging bay resource
      with charging_station.request() as request:
          yield request
          # charge car battery
          yield environment.timeout(charging_time)
```

FIGURE 3.10: Defining Electric Car Process

When the resource eventually becomes available the car begins service, which ends after the charging time elapses. The car process will pass the control flow back to the simulation once the last `yield` statement is reached.

Next the car arrival times and charging times must be specified before a simulation commences. This first illustration will use the fixed times in Figure 3.11, initially demonstrated in Section 1.6.1. Note that in a typical simulation with randomness, the times are generated within the simulation, and are not a priori constants (random times will be demonstrated next).

```
interarrival_times = [2, 8, 7, 2, 11, 3, 15, 9]
charging_times = [11, 8, 5, 8, 8, 5, 10, 12]
```

FIGURE 3.11: Defining Fixed Event Times

Next a loop to create all the events must be specified, which is shown in Figure 3.12. The code defines a fixed number of car processes to be created. Finally, the simulation is started after the car generation loop by calling the run method `environment.run()`. Note that the simulation duration may be specified in other ways such as a fixed ending time vs. the fixed number of car processes.

Running the program will produce the event output and summary statistics in Figure 3.13 (after adding some output print statements and statistics collection logic listed in Appendix B). The numerical results are identical to the manual calculations in Chapter 1.

Few changes are necessary to model random event times. After importing a random number generation library, the event generation loop simply replaces the fixed times with exponentially distributed random times per Figure 3.14. The provided val-

```
# simulate car processes
for i in range(7):
     arrival_time += interarrival_times[i]
     charging_time = charging_times[i]
     environment.process(electric_car(environment, '
            Car %d' % i, charging_station, arrival_time
                , charging_time))

environment.run()
```

FIGURE 3.12: Event Generation Loop and Simulation Start

ues are reciprocals of the distribution means for the interarrival and charging times respectively (i.e., the interarrival time mean = 6 and charging time mean = 5). The full Python code for these examples with additional print statements, data initializations, and statistics computation is listed in Appendix B.

3.3 AGENT BASED

Agent-based modeling uses autonomous decision-making entities called *agents* with rules of behavior that direct their interaction with each other and their *environment.* An agent must possess some degree of autonomy and be distinguishable from its environment. It must perform behaviors or tasks without direct external control reacting to its environment and other agents. Transitions move agents between *states* similar to how rates move entities between levels in continuous models.

Agents modeled may include many types of interacting individuals or groups such people, vehicles, robots, cells, data packets on a network, teams, organizations, and many others. Agents can represent entities that do not have a physical basis but are entities that perform tasks such as gathering information, or modeling the evolution of cooperation. Agent behavior can be reactive, e.g. changing state or taking action based on fixed rules, or adaptive after updating internal logic rules via learning.

Agents interact in environments, which are the spaces in which they behave. The environment may be discrete, continuous, combined discrete/continuous, or characterized by networks. Agent-based modeling can thus be combined with other simulation methods used in engineering sciences. For example, when simulating the interaction with the environment, the environment may be represented by a discrete or continuous field.

Agent-based simulations are a suitable tool to study complex systems with many interacting entities and non-linear interactions among them. Emergent behaviors can result, which are patterns generated by the interactions of the agents, which are often unexpected.

Agent-based modeling is relatively new without an extensive history of engineer-

```
Time  Event
2 Car 0 Arriving at station
2 Car 0 Entering a bay
10 Car 1 Arriving at station
13 Car 0 Charged and leaving
13 Car 1 Entering a bay
17 Car 2 Arriving at station
19 Car 3 Arriving at station
21 Car 1 Charged and leaving
21 Car 2 Entering a bay
26 Car 2 Charged and leaving
26 Car 3 Entering a bay
30 Car 4 Arriving at station
33 Car 5 Arriving at station
34 Car 3 Charged and leaving
34 Car 4 Entering a bay
42 Car 4 Charged and leaving
42 Car 5 Entering a bay
47 Car 5 Charged and leaving
48 Car 6 Arriving at station
48 Car 6 Entering a bay
58 Car 6 Charged and leaving

Summary
# of cars served = 7
Average waiting time: 3.9 minutes
Utilization: 0.948
Average queue length: 0.466
```

FIGURE 3.13: Program Output

ing usage like discrete event and continuous modeling. Its usage is increasing however. Agent-based models have been primarily developed for socio-economic systems simulating the interactions of autonomous agents (both individual or collective entities such as organizations or groups) to assess their effects on the system as a whole.

There are also many scientific and engineering applications to model groups of actors and their interactions based on behavioral rules. Urban systems in civil engineering to study people patterns, communities, vehicles, and traffic patterns are a natural fit. Example applications to-date in civil engineering include water resources [3], civil infrastructure [26], construction management [28], and more. They have been used to model complex healthcare systems and disaster response to simulate the movement of many individual people, and the consequent behavior of the crowd,

```
# simulate car processes
for i in range(7):
     arrival_time += random.expovariate(0.16)
     charging_time = random.expovariate(0.2)
     environment.process(electric_car(environment, '
             Car %d' % i, charging_station,
                arrival_time, charging_time))

environment.run()
```

FIGURE 3.14: Event Generation with Random Times

as they make their way out of the buildings and find transport to medical facilities.

Modeling dynamically interacting rule-based agents is ideal for transportation systems with emergent phenomena, when individual behavior is nonlinear and changes over time with fluctuations like traffic jams. It can be more natural to describe individual behavior activities compared to using processes in discrete event models.

Agent-based modeling is relevant for the engineering of systems of systems. Systems engineers can investigate alternative architectures and gain an understanding of the impact of the behaviors of individual systems on emergent behaviors. Other examples of engineering applications include supply chain optimization, logistics, distributed computing, and organizational and team behavior.

3.3.1 AGENT-BASED MODEL ELEMENTS

Agent-based modeling is typically implemented with an object-oriented programming where data and methods (operations) are encapsulated in objects that can manipulate their own data and interact with other objects. The behaviors and interactions of the agents may be formalized by equations, but more generally they may be specified through decision rules, such as if-then statements or logical operations.

Table 3.3 lists typical agent-based model elements. These definitions are consistent with the implementation in [8] and may vary slightly in other modeling tools.

TABLE 3.3: Agent-Based Model Elements

Element	Description
Population	An interacting group or collection of agents during a simulation described by the type of agents and the population size.
Agent	An individual actor in a population governed by its own rules of behavior.
State	An on/off switch for an agent. When the state is active, the agent is in that state and vice versa. One or more sets of states may be placed in an agent.
Transition	Transitions move agents between states in a model. When a transition is activated, an agent moves from one state to another. They are configured by what triggers the transition. Some different types of triggers may include a timeout, a probability, or a condition based on logical relationships to other agents or events.
Action	An action manipulates agents during a simulation. They are defined by what triggers the action and what the action does when triggered. An action can be to move an agent, to change a primitive's value, to add a connection to an agent, or other.

Example: Agent-Based Model for Car Behavior

Figure 3.15 shows an agent-based model for electric cars acting as interacting agents. This is a web-based simulation model using Insight Maker (see Appendix A). Consistent with the elements in Table 3.3, the population of cars is signified by the cloud at the top. Cars are defined as the agents that move between the states of *Waiting*, *Parked*, *Impatient*, and *Driving*.

This model is essentially a queuing process like the previous car charging example, but a variation with respect to the discrete event model is the consideration of cars leaving after waiting too long. Note the car behavior is actually dictated by a human agent inside and could alternatively be termed a driver rather than a vehicle.

Transitions are indicated by arrows in the model for moving between states. These are triggered by the decision to start looking for a charging space, the existence of an empty space, giving up after waiting too long, and completion of charging to

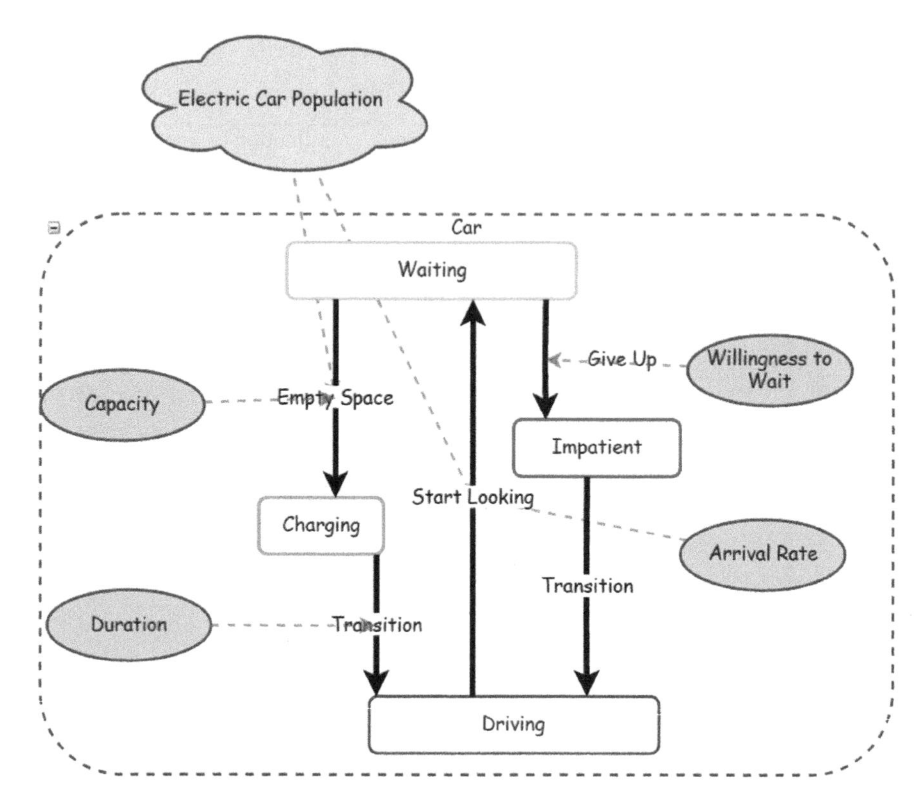

FIGURE 3.15: Example Agent-Based Model for Electric Car Charging

start driving again. The ovals are other model variables used in the logical rules of behavior. Dashed lines represent information links passing values.

A population of cars start off in a *Driving* state. At each cycle, according to a Poisson distribution defined by *Arrival Rate*, some cars transition to *Waiting* as they wait for an empty charging space according to:

```
RandPoisson(Round([Arrival Rate]))/Count(FindState([
    Electric Car Population], [Driving]))
```

The number of cars in the *Waiting* state represents a queue. If an empty bay is available for a car, then the state transitions to *Charging* per:

```
[Parking Capacity] > Count(FindState([Electric Car
    Population],[Charging]))
```

The cars remain in a *Charging* state according to a normal distribution and then transition back to *Driving* as they leave, modeled as:

```
RandNormal([Duration], [Duration]/2)
```

The model uses a timeout for the impatience behavior. If a car is in the *Waiting* state for a period longer than *Willingness to Wait*, then the state times out and transitions to *Impatient* and immediately transitions to *Driving* again. Figures 3.16 and 3.17 show a simulation timeline and state transition map for a simulation run.

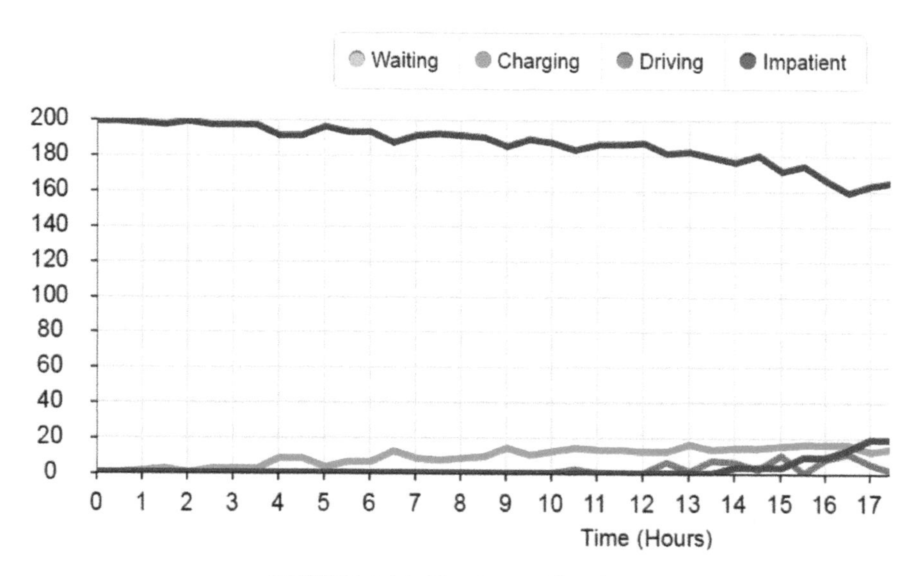

FIGURE 3.16: Simulation Timeline

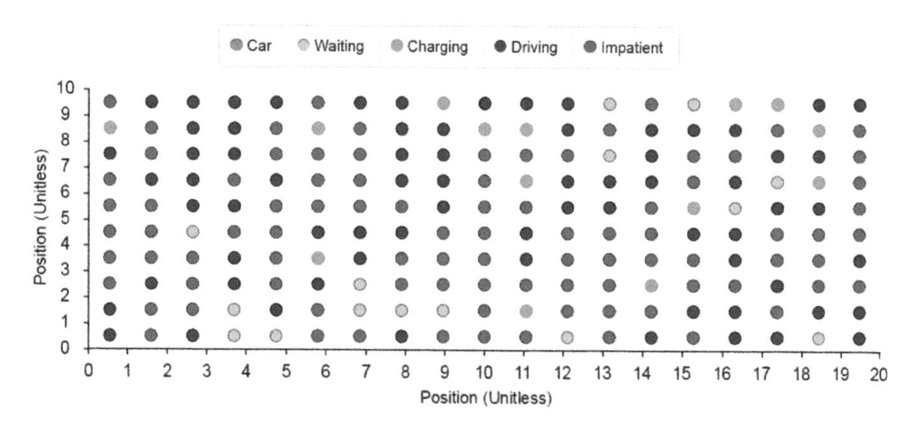

FIGURE 3.17: State Transition Map

3.4 SUMMARY

Systems and/or their models can be discrete, continuous, or a combination. All classes of systems may be represented by any of the model types. In continuous models the variables change continuously with respect to time. Time advancement is in equal small increments, with all variables recomputed at each time step. Discrete system models are event based, whereby an event calendar is used for timing. State changes occur in discrete systems at aperiodic times depending on the event times. Time advancement is from one event to the next, with no changes in variables between events.

Before choosing a modeling method for engineering applications, an important question that should be answered is, "What is continuous and what is truly discrete?" Are there discrete aspects that need to be preserved for the purpose of the study or is an aggregate view sufficient?

System dynamics is the most widely used form of continuous simulation. In the continuous view, process entities are represented as aggregated flows over time. The approach does not track individual events; rather, flowing entities are treated in the aggregate and systems can be described through differential equations. System dynamics provides a rich modeling environment incorporating many types of formulations for this.

System dynamics model elements are levels, sources and sinks, rates (flows), information links, and auxiliary variables. Levels represent real-world accumulations and serve as the state variables describing a system at any point in time. Rates are the flows over time that affect the levels. Rates may represent decisions or policy statements. Rates are computed as a function of levels, constants, and auxiliaries. Sources and sinks indicate that flows come from or go to somewhere external to the system. Models are built using conserved flow chains where the levels store the flowing entities.

Discrete event models consist of flows of individual entities, characterized by the values of their respective attributes. The characteristics of the entities can change during a simulation run. These changes occur instantaneously as the simulated time lapses. The entities move through a system represented as a network of nodes, perform activities by using resources, and create events that change the state of a system. The next-event time-advance approach is the standard procedure used for computing in discrete event modeling.

Discrete event simulation represents system behavior using a set of common constructs that correspond well with intrinsic, measurable, real-world phenomena. Standard discrete processes include entity creation, entity movement and traversal, entity use of resources, entity accumulation (queues), and entity termination. In tools they are used to represent the movement and accumulation of entities along paths through system nodes. Modeling tools also provide auxiliary operations and statistics specification for measuring and assessing system performance data.

Agent-based models contain agents that are autonomous decision-making entities with rules of behavior that direct their interaction with each other and their environment. Agents may include many types of interacting individuals or groups. Agents interact in environments, which are the spaces in which they behave. The environment may be discrete, continuous, combined discrete/continuous, or characterized by networks.

Agent-based model elements include agents, their populations, states, transitions, and actions. Individual agents are governed by their own rules of behavior and populations store agents during a simulation. Agents have internal states, and transitions are used to move agents between states in a model. Actions manipulate agents during a simulation and are defined with event triggers.

This chapter has demonstrated simple, illustrative models for continuous, discrete event and agent-based types of modeling and simulation. The reader is encouraged to look further in the references and upcoming chapters, see Appendix B for modeling tools, experiment with the tools and the provided models in this book.

4 Randomness

4.1 REPRESENTING UNCERTAIN VALUES

A simulation can be deterministic, stochastic, or mixed. In the deterministic case with no stochastic parameters, inputs are specified as constant values and uncertainty in the values is not addressed. In a purely deterministic model, only one simulation run is needed for a given set of parameters.

Stochastic modeling recognizes the inherent uncertainty in parameters and relationships. For many modeling purposes, the stochastic nature of a system is an essential feature to be represented. It would be risky to draw conclusions based on the results of a single simulation run when it is one of many possibilities. Rather than using a point estimate, a stochastic variable takes random values drawn from a specified probability distribution. (The mathematical notation in this chapter and throughout the book denotes random variables by capital letters such as X and Y. The values that the random variable take on are denoted by lowercase letters such as x and y.)

To demonstrate the relationship between stochastic variables and probability distributions, consider that when inputs cannot be characterized with certainty, as point estimates, they can be characterized with some measure of uncertainty. The uncertainty has a distribution form. For example, uncertain values of the effort required to perform an activity usually have a minimum effort and a maximum effort. If that is all we know about the effort required for the activity, we might assume that the actual effort can be any value between the minimum and maximum values with equal probability. With this assumption that values between the minimum and maximum are equally likely, the values are said to be uniformly distributed. Suppose that in addition to a minimum and a maximum, we know that a value between the two extremes is the most likely value for the activity's effort. With that additional knowledge we may assume that the distribution is triangular, not uniform.

Simulation provides a flexible and useful mechanism for representing uncertainty in complex systems. It requires however, that bounds on the uncertainty of parameter values be understood and evaluated. Simulation propagates uncertainties through a model, allowing the modeler to see the effects of uncertain inputs on outcome variables. Furthermore, stochastic and mixed models can be run many times using different random input values to produce different output values for each run. The output values can then be analyzed statistically across simulation runs and be used to assess risk in terms of probabilities See Section 4.4 on Monte Carlo analysis for details on generating the random inputs, and Chapter 6 on statistical analysis of the outputs.

This chapter discusses the concepts just mentioned: randomness as the underlying construct for probability distributions, basic concepts of probability distributions with some common probability distributions, and use of probability distributions in

Monte Carlo trials. More comprehensive treatments of these topics can be found in [17] and [16] (the latter is oriented toward discrete systems).

4.2 GENERATING RANDOM VALUES

Randomness is essential to simulations having non-deterministic inputs. If input values are not random, then they are biased, leading to biased outputs. Thus, randomness is an assurance that uncertainty is fairly represented.

Probability distributions are based on independent, random values. For example, we might say that an input parameter is "normally distributed with mean x and variance v." In modeling such a distribution, we have to be able to produce random values that are normally distributed with mean x and variance v. Several methods for generating random values are available and the interested reader can consult Law [17] for detailed discussions of these methods.

The most common generation method is *inverse transform*. It is discussed in Section 4.4.1 to show how a variety of probability distributions can be generated from the output of a random number generator that produces random value uniformly between 0 and 1, denoted $U(0,1)$. Thus for many probability distributions, the problem of generating values is reduced to producing $U(0,1)$ values.

Generating a stream of values randomly is challenging. Fortunately, there are algorithms that generate streams of numbers that appear to be random numbers. These algorithmic number stream generators are called pseudorandom number generators (PRNG) because the number streams are not truly random, but appear sufficiently random to serve the purpose of behaving randomly. The quality of PRNGs varies widely. The quality of such a generator is judged by the unpredictability of the values it produces. A number of tests have been developed for assessing the quality of a PRNG based on its output. These will be described briefly in the next section.

The simplest and most common of these algorithms is called a linear congruential generator (LCG). LCGs take the form of

$$X_{n+1} = (aX_n + c) \mod m \tag{4.1}$$

where
a is a coefficient
c is an increment
m is the modulus
X_0 is a starting value, or seed

All four values are integers and their choices determine the LCG. To obtain $U(0,1)$, $U_n = Xn/m$. For example, if $a = 3$, $c = 8$, $m = 27$, and $X_0 = 10$, X_1 is the remainder of $(3*10+8) \mod 27 = 11$ (the remainder of dividing 38 by 27) and $U_1 = 0.407$. Continuing, X_2 is $(3*11+8) \mod 27 = 14$, and $U_2 = 0.518$.

This example illustrates the fact that PRGNs are deterministic and therefore, not truly random. Thus, they are pseudorandom number generators. However, the determinism of PRNGs is a desirable trait because it allows us to repeat a simulation run and obtain the same results.

Another property of LCGs is that they can take on values from 0 to $(m-1)/m$. When a value is repeated, the LCG repeats the series. The length of the series is the period of the generator. With good choices of large values for the four integer variables, very long periods can be produced. Long periods are desirable because a simulation model may use millions of random values.

Finally, the example illustrates that the pseudorandom number stream is dependent on the assigned values. If one tried different values for the LCG, one will find that the number streams produced by each LCG vary in their quality.

The quality of a PRNG is its ability to produce streams of numbers that are apparently random. PRNG quality can be tested by testing the number stream it produces. Number streams can be tested for independence and for uniformity. One test for uniformity is to form a histogram of the numbers and see whether each interval contains approximately the same number of values.

A *runs test* can be used to test a number stream for independence. In this test, a run is a sequence of numbers of increasing value. The test counts the number of runs of length i for $1 \leq i \leq 5$ and $i \leq 6$. Law [17] provides the algorithm for calculating the test statistic that is used to decide independence of the values in the stream.

Simulation software programs generally provide algorithms for generating random numbers in a wide selection of probability distributions. Although the PRNG used in a program may be proprietary, it is often based on a well-known PRNG and the vendor may disclose from which published PRNG their generator is adapted. If so, the modeler may have more confidence in the generator's quality. However, a modeler can generate a stream of thousands of values and then test these values for uniformity and independence to check the quality of a simulation program's PRNG.

Another option for obtaining a random number stream is to use a stream obtained from an external source, store it, and use it in a simulator. A number of websites offer random number generation. For example, `http://random.org` generates random numbers from atmospheric noise.

4.3 PROBABILITY DISTRIBUTIONS

Probability is a quantitative measure of the chance or likelihood of an event occurring. A probability distribution is an arrangement of data that shows the probability or frequency of measurements vs. measurement values (also called a frequency distribution with measurement counts in discrete bins). See Figure 4.1 for a general probability distribution function. The graph is also known as the probability density function $f(x)$ of the random variable x_i. The total area under a *probability distribution function* (PDF) is unity. Distributions may be continuous or they can be discrete. The PDF may also go by the name *probability density function.*

Frequency histograms of data values are also probability distributions. Intervals are used on the abscissa to show the number of measurements within discrete bins, and this bar chart depiction is called a histogram seen in Figure 4.4. These probability distributions are used to characterize random input data and the probabilistic results of simulations, such as when using the Monte Carlo simulation approach.

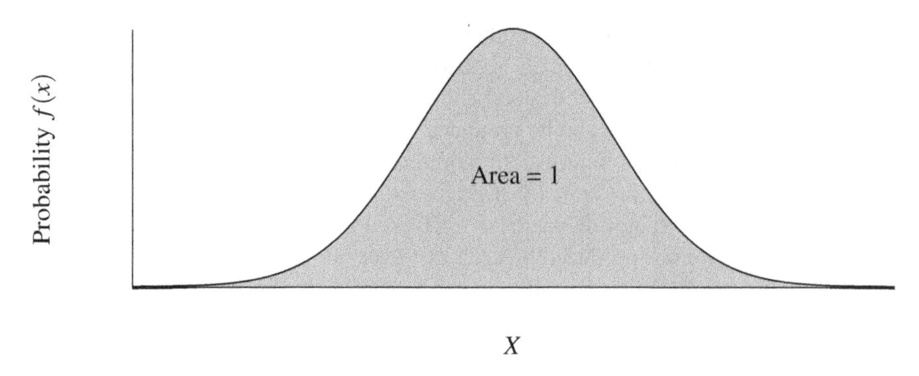

FIGURE 4.1: Probability Distribution Function (PDF)

The integral, or cumulative form of the PDF is called the cumulative distribution function (CDF). It is also sometimes called a probability distribution function, so we will use the term CDF to distinguish it. The CDF is defined as $F(X)$, the integral of $f(x)$:

$F(X) = \int f(x)dx$

The CDF has the following properties:

$0 \leq F(x) \leq 1$ for all x
$F(x)$ is nondecreasing.

A CDF is shown in Figure 4.2.

The CDF gives the probability that a random value x_i is less than or equal to x. This probability is expressed as

$P(x_i < x)$.

The final value of a CDF is always one, since the entire area under the PDF is also one. For discrete functions, $F(X)$ is the summation of $f(x)$ and holds the same properties. See Figure 4.3 for an example discrete PDF. Generally we can use continuous probability distributions to also model discrete quantities, so this section will focus on continuous distributions.

Figure 4.5 shows a general cumulative distribution corresponding to the histogram in Figure 4.4. The PDF is the first derivative of the CDF, since the CDF is the integral of the probability distribution function $f(x)$. A common application of the cumulative distribution is to generate random variates for the Monte Carlo method, and is described in Section 4.4.

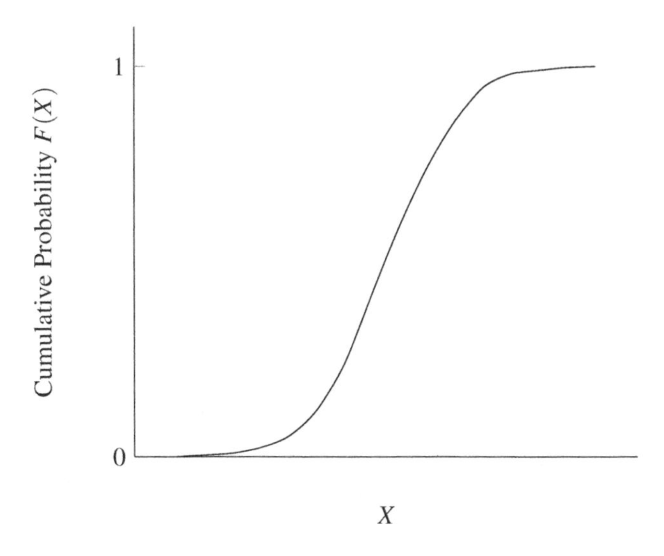

FIGURE 4.2: Cumulative Distribution Function (CDF)

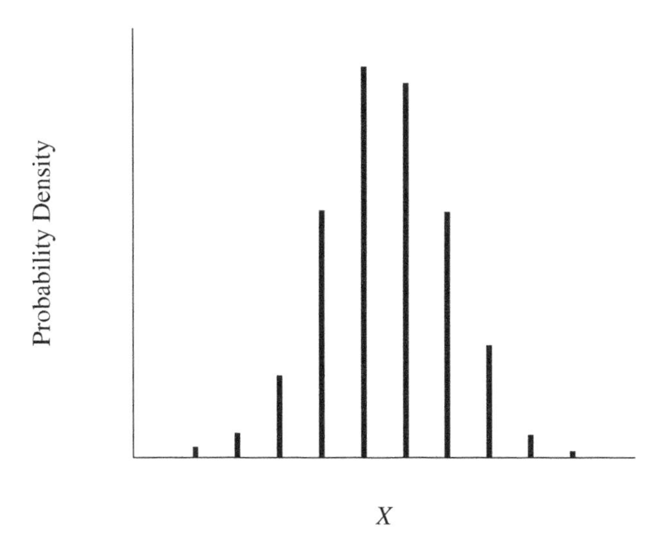

FIGURE 4.3: Discrete Probability Function

4.3.1 INTERPRETING PROBABILITY DISTRIBUTIONS

Recall that the total area under a PDF is unity. The area under a portion of the PDF is the probability of a random measurement x_i lying between the two corresponding values x and $x+dx$. In a symmetric normal distribution, 50% of the values lie below

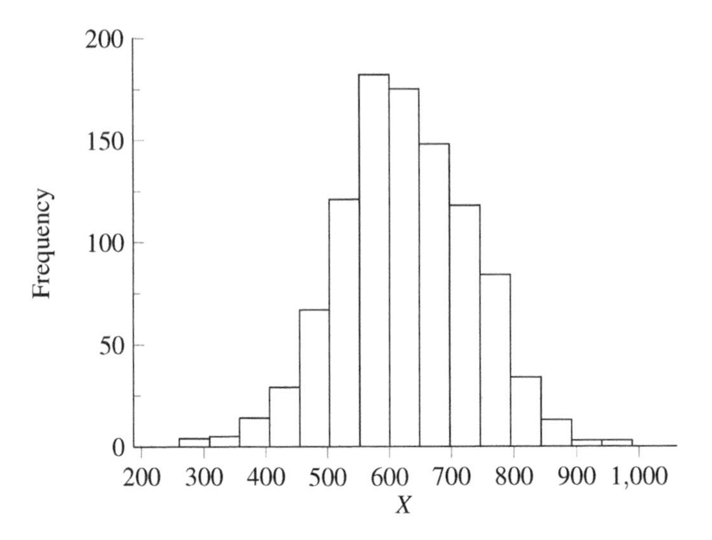

FIGURE 4.4: Frequency Histogram

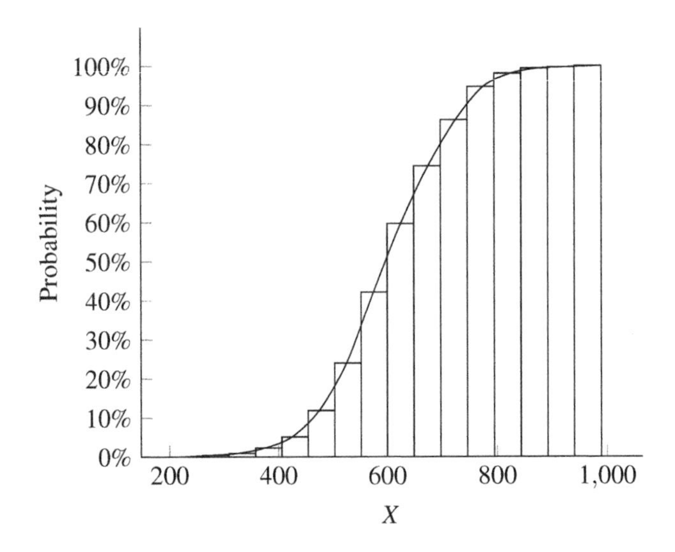

FIGURE 4.5: Cumulative Distribution Function from Histogram

the mean and 50% above.

For example, the shaded area in Figure 4.6 corresponds to the probability of waiting time being between 2.6 and 2.8. This probability can be calculated from the corresponding cumulative distribution in Figure 4.7. It equals the difference of cumulative probabilities evaluated at 2.8 and 3.0, or $F(3.0) - F(2.8) = .5 - .2 = .3$. Thus there is a 30% chance of waiting time lying between 2.8 and 3.0.

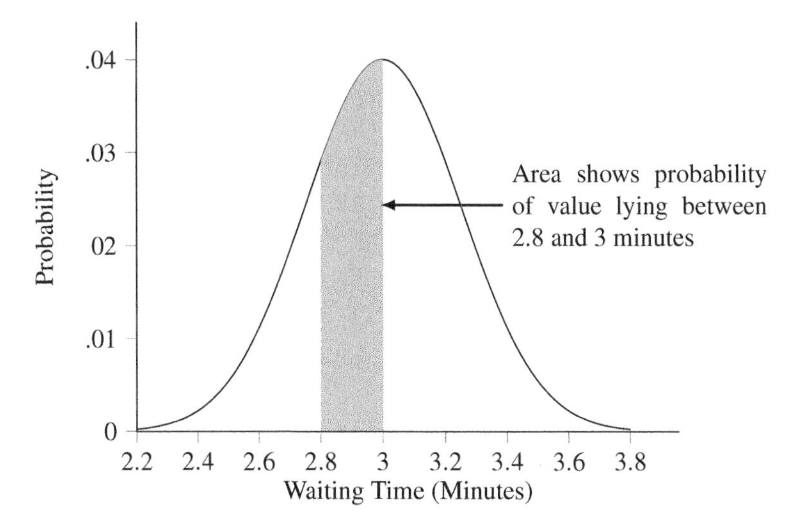

FIGURE 4.6: Interpreting Probability Region with Distribution

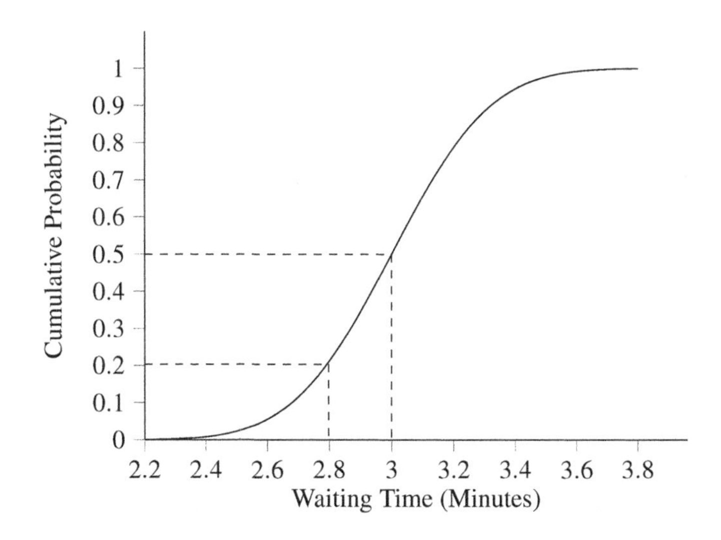

FIGURE 4.7: Calculating Probability with Cumulative Distribution

The cumulative form of a resulting output distribution can also be used as a confidence level chart. Confidence level refers to the probability of not exceeding a given value. For example, the CDF in Figure 4.7 can be used to determine the confidence level as the cumulative probability for a given value. This is done by finding the corresponding probability on the y-axis for a specific value on the x-axis. For example, there is a 50% chance that the waiting time will be less than or equal to 3 minutes as visualized on Figure 4.7.

A CDF can also be transposed for ascertaining confidence levels. A sample confidence level chart in the form of a transposed CDF is shown in Figure 4.8. The figure represents the results of running many simulation runs, and can be used to assess cost risk. For example, there is a 70% chance that the simulated project will cost $100K or less.

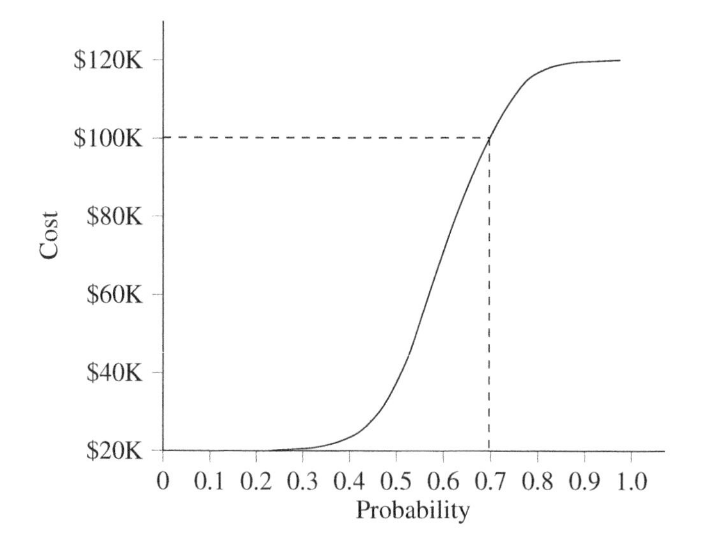

FIGURE 4.8: Confidence Level Chart Using Transposed Cumulative Probability Distribution

4.3.2 COMMON PROBABILITY DISTRIBUTIONS

There are a very large number of probability distributions used in different fields. This section provides an overview of some of the simpler and most commonly used distributions that can be applied to engineering applications. The reader may consult [17] for more probability distributions and more information on each distribution.

Uniform

A uniform distribution represents a range of equally likely outcomes with an equal number of measures in each interval. As shown in Figure 4.9, all values lying between the range endpoints of a and b are equally likely.

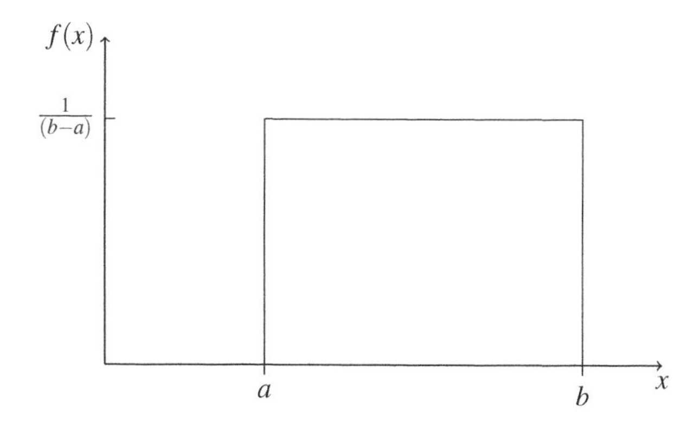

FIGURE 4.9: Uniform Probability Distribution

It can be used as an initial model for a quantity that is thought to be randomly varying between a and b but about which little else is known. It can be used in the absence of distribution data, or to represent equally likely outcomes (e.g. decision choices).

The uniform distribution $U(0,1)$ is also essential for generating random values for all other distributions as mentioned earlier. See more details in Section 4.4 on how it enables the inverse transform for the Monte Carlo approach.

Triangular

The triangular distribution has three parameters: the minimum value (a), the maximum value (b), and the mode (c) value. It can be used when no recorded data is available but estimates of the minimum, maximum, and most likely values of a random variable are available. It is easy to understand because it is intuitive.

It allows for skewedness and the limiting cases occur when either $c = a$ (left triangular) or $c = b$ (right triangular). This distribution is useful when a random distribution is known to have a mode but the shape of the distribution is otherwise unknown. It is commonly used to express technical uncertainty because of its use of pessimistic, most likely, and optimistic inputs.

Normal

A normal distribution is symmetric around the mean and bell shaped (also called a bell curve) in which most measures occur closest to the center of the distribution and fewer in the intervals farther from the center. Figure 4.11 shows a normal distribution centered around its mean μ.

The symmetric normal distribution is used for outcomes likely to occur on either side of the average value. The width of the distribution, or its dispersion, varies as expressed in the standard deviation σ. Figure 4.12 shows normal probability distri-

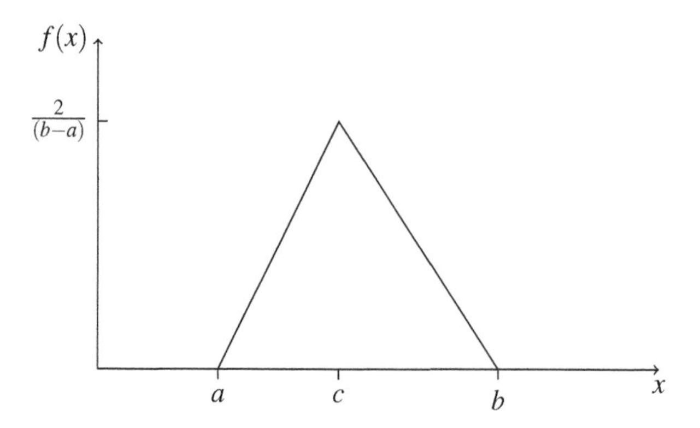

FIGURE 4.10: Triangular Probability Distribution

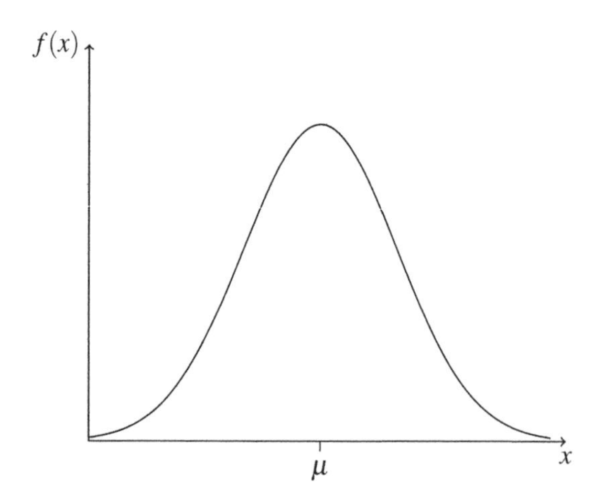

FIGURE 4.11: Normal Probability Distribution

butions with different standard deviations. All distributions in the figure have a zero mean but the dispersions vary as denoted in the figure.

Lognormal

The lognormal is a continuous distribution positively skewed with a limitless upper bound and known lower bound. It is skewed to the right to reflect a tendency toward higher values. The lognormal distribution can occur for phenomena that exhibit the same kind of variation seen in a normal distribution but have a lower bound such as zero. For example, waiting times in a queue can be lognormally distributed.

In the skewed distribution the measurements cluster around a certain value, but

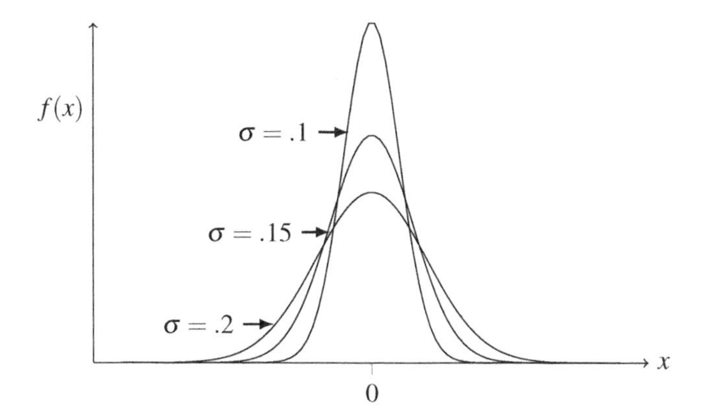

FIGURE 4.12: Normal Distributions with Varying Standard Deviations (Mean = Zero)

that value is not in the center of the distribution. In the lognormal distribution, $ln(x)$ is normally distributed. The distribution mean is the mean of $ln(x)$, and the standard deviation is the standard deviation of $ln(x)$. Figure 4.13 shows a sample lognormal distribution. The dispersion around the mean is again characterized by the standard deviation.

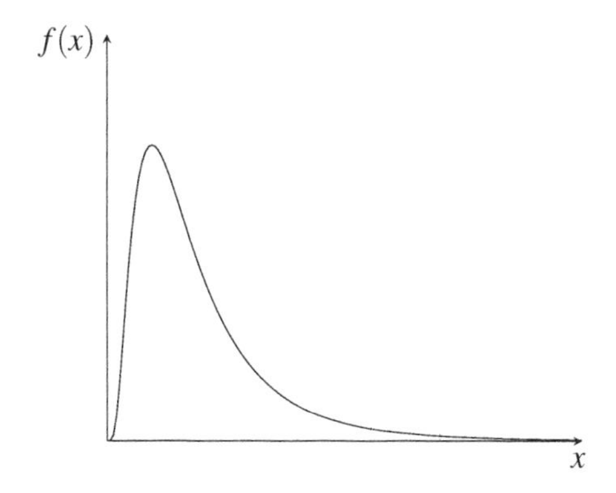

FIGURE 4.13: Lognormal Probability Distribution

The lognormal distribution is appropriate for modeling parameters with long right tails in their distributions. For example, it is useful for modeling the overall size or complexity of an engineering task because one tends to underestimate the maximum possible values.

One drawback of a lognormal function is that it has an extremely long tail, which may be unrealistic. For this, a right-truncated lognormal function bounds the distribution to more realistic end values. The left-truncated lognormal avoids non-positive values or other inconsistent values in a lognormal (e.g. when the mean is close enough to zero that the spread goes negative).

The lognormal function can take on shapes similar to the gamma function or the Weibull distribution (see Chapter 5). The lognormal is also used to characterize uncertainty for intermediate modeling relationships vs. parameter values.

PERT

The PERT probability distribution,which is used in many modeling and project scheduling tools, is a simplified form of a Beta distribution. It is a rounded version of the triangular distribution that can be skewed or resemble a normal distribution. It is specified by three parameters: a minimum value, most likely, and a maximum value as shown in Figure 4.14. Typically, the minimum and maximum represent 5% and 95% cumulative probabilities respectively.

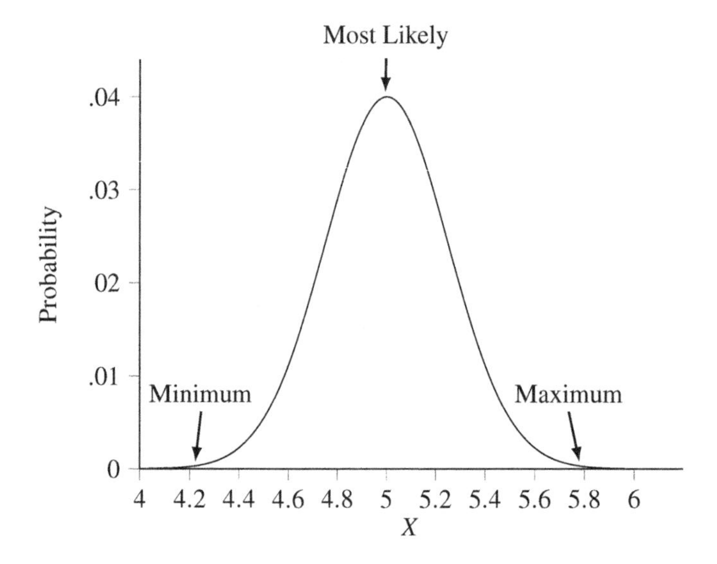

FIGURE 4.14: PERT Distribution and Its Parameters

Gamma

Another skewed distribution is the gamma function. It is shown in Figure 4.15 with varying values of the beta parameter while holding alpha constant. Its shape can be used to model parameters with varying degrees of a right tail. The gamma function, the lognormal, and the Weibull distribution can take on very similar shapes. Also see

Section 5.4 for a Weibull example of this type of distribution and discussion of its properties.

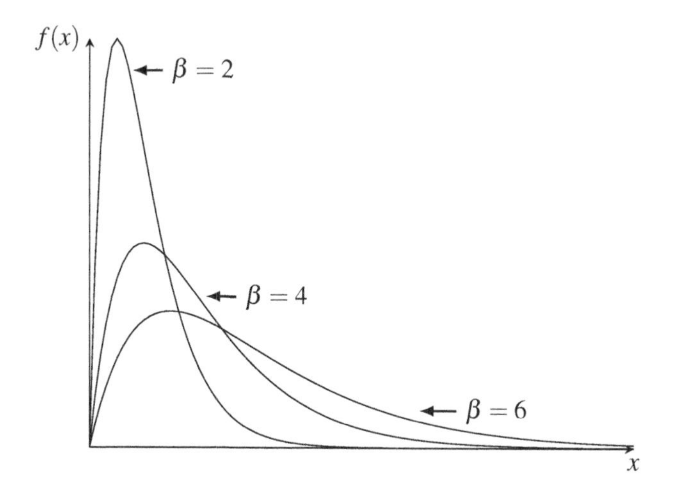

FIGURE 4.15: Gamma Probability Distributions

Empirical Distributions

When actual observed data is available, an empirical distribution can be derived from it instead of resorting to a theoretical distribution. Sometimes a theoretical distribution won't adequately fit the data, and often an empirical distribution will be more realistic since it is predicated on real data (assuming there is enough data).

For example, some data might exhibit a bimodal distribution and the best way to represent it is building up your own empirical distribution. Construction of empirical distributions is discussed in Section 5.4.4.

4.3.3 SUMMARY AND USAGE OF PROBABILITY DISTRIBUTIONS

A visual summary of continuous and discrete probability distributions with their general equations is shown in Figure 4.16. The PDFs display the probability density, which may also be interpreted as frequency. The CDFs integrate the PDFs across the x-axis to obtain the cumulative probabilities.

What are the best probability distributions to use? The uniform distribution is a simple one to start out with, and is relevant when there is an equal probability of a variable within a given range or no better information is known. Next in sophistication is the triangular distribution. It doesn't have to be symmetric, and some skewness can be modeled with it.

A normal distribution can be used to model some quantities more realistically than a triangular distribution. But it is a symmetric distribution, and care should be

Continuous Distributions

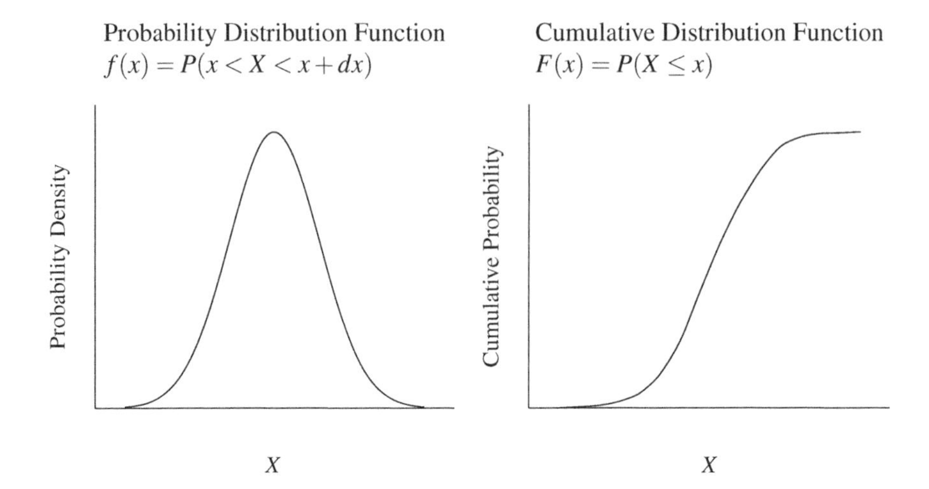

Discrete Distributions

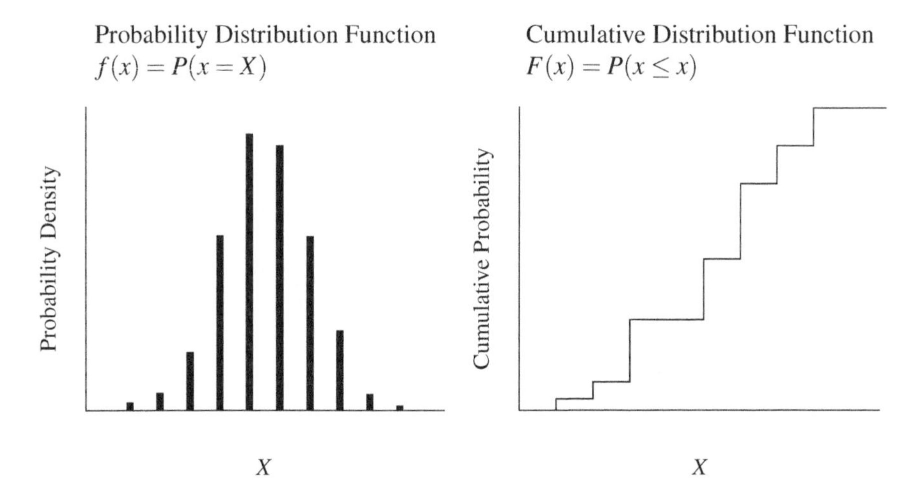

FIGURE 4.16: Summary of Probability Distributions

used when using it to ensure that the quantity being modeled is indeed symmetric.

The PERT distribution is simple to implement and can approximate a symmetrical normal distribution or a skewed distribution like the triangle or lognormal. It is available in many packages and often gives good enough results given the inherent uncertainties. But if more precise information is known about a parameter distribution, then one should consider an alternative if the PERT is not a good fit and modeling precision is important.

Positively skewed distributions with long tails are recommended to be realistic for many situations to compensate for optimistic tendencies when modeling parameters that are related to cost, effort, or complexity. The asymmetric lognormal distribution is relevant for modeling the time to perform some task, particularly since scope is usually underestimated and unconsidered details tend to increase the job size. It will account for the asymmetry in the expected outcomes. It is often useful for quantities that are the product of a large number of other quantities (by virtue of the Central Limit Theorem described in Chapter 6). See Chapter 5 for further guidance on choosing probability distributions, Chapter 6 on statistical analysis of random outputs, and Chapter 7 for an extended case study applying randomness concepts.

4.4 MONTE CARLO ANALYSIS

Monte Carlo is a "game of chance" technique used to solve many types of problems by applying random sampling instead of analytic methods. For a model with a number of input parameters each having probability distributions, the distributions are independently and randomly sampled producing various values for each output parameter. Using these for a large number of simulation runs, the output parameter values will be characterized as random distributions. This method of analysis is often referred to as Monte Carlo simulation and it does not necessarily imply time-based simulation. So the term "Monte Carlo" is used broadly to refer to any random sampling method.

The following steps are performed for n iterations in a Monte Carlo analysis, where an iteration refers to a single simulation run:

1. For each random variable, take a sample from its probability distribution function and calculate its value.
2. Run a simulation using the random input samples and compute the corresponding simulation outputs.
3. Repeat the above steps until n simulation runs are performed.
4. Determine the output probability distributions of selected dependent variables using the n values from the runs.

4.4.1 INVERSE TRANSFORM

Monte Carlo uses random numbers to sample from known probability distributions to determine specific outcomes. The inverse transform technique for generating random variates is convenient for this. First a random number r is chosen that is uniformly distributed between 0 and 1, denoted as $U(0,1)$. It is set equal to the cumulative distribution, $F(x) = r$, and x is solved for. A particular value r_i gives a value x_i, which is a particular sample value of X. It can be expressed per Equation 4.2:

$$x_i = F^{-1}(r_i) \tag{4.2}$$

This construction is shown in Figure 4.17 to generate values x_i of the random variable X that represents service time. It graphically demonstrates how the inverse transform uses the cumulative probability function for generating random variates using random samples between 0 and 1 for input.

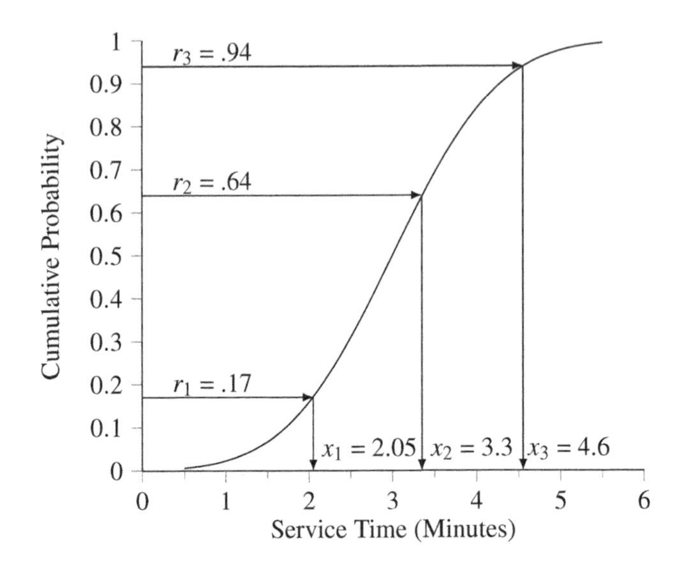

FIGURE 4.17: Cumulative Probability Distribution and Inverse Transform for Three Random Draws

Referring to Figure 4.17, first a $U(0,1)$ is generated, then the cumulative distribution is used to find the corresponding random variate. In the example from a normal distribution, the first random number of $r = .17$ generates a value of $x = 2.05$, the second random draw of $r = .64$ produces a value of $x = 3.3$, and the third draw of $r = .94$ generates $x = 4.6$ per the figure.

There is a strong intuitive appeal to the inverse transform method. The technique works because more random number inputs will hit the steep parts of the CDF, thus concentrating the random variates under those regions where the CDF is steep: exactly the same regions where the PDF is high. Since the PDF is the derivative of the CDF ($f(x) = F'(x)$), $f(x)$ can be viewed as the slope function of $F(x)$. Thus, the CDF rises most steeply for values of x where $f(x)$ is large and conversely it is flat where $f(x)$ is small.

The Monte Carlo method will fill out a distribution of random variates prescribed by the PDF, since most of them will be from the largest areas of the original distribution. The corresponding distribution of random variates will thus better resemble the actual PDF with more iterations.

4.5 SUMMARY

The random effects inherent in systems should be modeled to build representative simulations. Probability distributions are used to express the stochastic ranges of simulation inputs and account for the uncertainties. Randomness is also important to consider in simulation output analysis to analyze results in terms of probabilistic ranges that represent uncertainty. Thus the inputs and outputs are best treated as probability distributions. When a probability distribution is used to express the output of several simulation runs, its cumulative form can be used to assess the confidence level for a given output value.

Probability distributions are based on independent, random values. The total area under a probability distribution function (PDF) is unity. Distributions may be continuous or they can be discrete. Frequency histograms of data values are also probability distributions. The integral, or cumulative form of the PDF is called the cumulative distribution function (CDF). A common application of the cumulative distribution is to generate random variates using the inverse transform method.

Common probability distributions for modeling systems include the following simple ones:

Uniform distribution represents a range of equally likely values.
Triangular distribution is specified with a minimum, maximum, and most likely value.
Normal distribution is a bell-shaped symmetric distribution defined by its mean and standard deviation.
Lognormal distribution is a positively skewed continuous distribution.
Gamma distribution is right-skewed and can take on various shapes defined by its constants.

Empirical distributions can also be derived when actual observed data is available, instead of using theoretical distributions.

The Monte Carlo method is a convenient way to create random variates for simulation runs through the use of probability distributions and inverse transforms. Uniform random numbers between 0 and 1 are used to index into a known cumulative probability distribution (as an inverse transform) to generate random variates matching the distribution. These generated random variates are then used as inputs to drive a simulation. Multiple input parameters can be randomized to reflect the system uncertainties. Most modeling tools support Monte Carlo simulation. They provide features for generating random numbers from a wide selection of probability distributions to feed Monte Carlo simulation.

The discussion of randomness in this chapter is an essential basis for many of the topics regarding model input and output analysis in the following chapters.

5 Simulation Input Analysis

The previous chapter described the role of randomness in representing uncertainty when using a model to describe a system. This chapter describes some methods for analyzing data using the probability concepts described in Chapter 4. To build a representative simulation model that accounts for uncertain parameter values, input parameters are modeled as values that are randomly distributed. For example, data may indicate that a value of an activity duration, such as electric car charging, is most likely 60 minutes but no less than 40 minutes and no more than 120 minutes. These values indicate that a triangular distribution can be used to model the charging duration, and that every time we need a value for the duration we can randomly sample the triangular distribution to obtain a value.

In constructing a model, one may realize that many elements in a model must be characterized with an input that exhibits some degree of uncertainty. These can include entity arrival times, number of entities arriving, activity durations, number of resources in a pool, capacity of an activity, routing of entities for processing, the cost of a resource, proportion of entities having a specific characteristic, and so forth. Every model input should be considered for its uncertainty. The overall steps performed during the input analysis described next are as follows:

Identify input variables.
Decide on types of input values.
Collect and/or elicit data on the input variables.
Estimate the parameters of a data distribution.
Fit known distributions to data, test fits, and select a distribution.

5.1 IDENTIFYING INPUT VARIABLES

Chapter 2 addressed model scope in terms of specifying a boundary for a model. A boundary is defined by specification of inputs and outputs. Therefore, choices of input variables are based on model scope and help define scope. When choosing inputs, one must consider whether an input incorporates enough scope to fulfill the modeling purpose. For example, a model of on-orbit satellite availability might start with rocket production if the purpose of the model is to include the effects of production timing. But if an adequate supply of satellites is assumed, the model can start with launch events. If the question for the satellite availability model is "What production rate is necessary to maintain 90% availability of two satellites?" then production variables such as production period, production capacity, storage capacity, and launch rate must be input. Alternatively, if the question is "What launch rate is necessary to maintain 90% availability of two satellites?" then production variables are outside the model scope but launch rate is in the scope.

The level of abstraction of a model also influences choices of input variables.

For example, if a satellite availability model is used to understand launch rate requirements, then the launch frequency required to provide the required availability is sufficient. But if the model is used to understand launch rate feasibility for meeting the availability requirement, then necessary inputs may include number of launch pads, launch processing time, and historical delays due to weather. As more detail is required to fulfill a modeling purpose, more input variables are introduced and the work of input analysis can grow. Therefore, the number of input variables should be kept to what is essential for the model purpose, scope, and level of abstraction.

5.2 DECIDING TYPES OF INPUT VALUES

Input values can be either constants or random distributions. Some variables can reasonably be expected to take constant input values for a series of runs, either because they represent a current unchanging status, or because they are a matter of policy. For example, the number of launch pads available on a launch range may be three and when the model is run to represent current launching capacity, this input is always three. It may be set to four when representing a future capacity. Another variable, such as the minimum time between launches, which represents a safety policy, can also be treated as a constant.

Another example is a delay. Delays are variables that are treated as either constant or randomly distributed, depending on the parameter and level of abstraction. In system dynamics modeling, constants are often used as inputs to delays due to the high level of abstraction. For example, the delay in converting a test bed from testing one product line to testing another product line can be modeled as a second-order delay requiring two weeks for conversion. In a discrete event model, or a model showing more detail, the same variable might be represented as a minimum required time for conversion plus a risk of delay due to unforeseen problems, such as late delivery of required hardware. In this case, the parameter value is the sum of a constant and a randomly generated value.

Characterizing the values of an input parameter as a random variable requires more much work than using a constant value. The additional work is in obtaining and analyzing data, fitting a distribution, and implementing and testing values. Therefore, care should be taken in specifying how many parameters and which ones are to be random variables. The level of abstraction may not require the variation expressed in a random variable. Or, the variable may be not be influential enough to warrant producing a random distribution. When in doubt, one can start with a constant and test the model with high and low constant values to see whether model outputs are affected enough to warrant treating the input parameter as a random distribution. In the end, the relative influence of all inputs can be calculated in a sensitivity analysis exercise that uses a designed experiment to vary all input values. This will be discussed further in the chapter on output analysis.

5.3 COLLECTING DATA ON INPUT VARIABLES

5.3.1 COLLECTING DATA

Data may or may not have been collected for a parameter. When data has not been collected for a model parameter, then a decision must be made about collecting data. Generally, data collection is expensive and time-consuming. Data collection can easily become the task requiring the longest duration in a modeling project. Whether waiting on stakeholders to obtain data and gain permission to release it to modelers, or going through a data measurement and recording process, data collection can take a long time. Therefore, it is not surprising that data for small models needed within a short time (such as a few weeks) are often elicited from experts rather than collected through measurement systems.

5.3.2 ELICITING DATA

Eliciting data is necessary when no data exists for a model parameter and the cost or duration of data collection is prohibitive for the sponsor's purpose. These factors, lack of data, and potential benefit of using expert information, are often present when modeling organizational processes. In these cases, the modeler must first ask whether the data that can be elicited from domain experts will be adequate for the modeling purpose. If the modeler and sponsor recognize the value of producing a model using data provided from the experience of experts, then the modeler can proceed with confidence that the effort will be beneficial.

Data elicitation should be treated as a structured interview of multiple experts who make independent judgments about parameter values and then compare them with the intention of developing a consensus. The Delphi method [11] is such an approach for forecasting, but is easily adapted to eliciting actual values under the constraints of time and expert availability. Another very useful resource for structuring elicitation of values is Hubbard's work on Applied Information Economics [12].

When consulting experts in a system or process, they can be asked for a most likely value of a parameter, a minimum value, and a maximum value. The definition of minimum and maximum can lead to debates as to whether these mean extreme values (the lowest and highest values either observed or possible) or "90% values." One way to facilitate this discussion is to ask for low and high values as usually occurring and then ask whether extremes beyond the usual lows and highs can appear. The following series of questions is helpful for eliciting data for each parameter.

> What is the lowest value of the parameter most of the time?
> What is the highest value of the parameter most of the time?
> Is a value between these two more likely than either one? If so, what is it?
> If the parameter has a most likely value, can values occur that are less than the lowest value or more than the highest value?

The answers to the first three questions indicate whether the distribution is uniform or triangular. Uniform distributions are used in the absence of any information except minimum and maximum values. When a most likely value can be stated, a triangular distribution is indicated. If a triangular distribution is used, the minimum and maximum values should cover all of the possible values, not most of them. However, if a most likely value is given, and the answer to the fourth question above indicates that extreme values can occur, then a continuous distribution with a tail is indicated. Although a distribution can be two-tailed, like a normal distribution, experts may indicate extreme values are more likely on one side, suggesting a skewed distribution.

When eliciting parameter values, note the degree of conviction or assurance with which responses are offered. If an expert offers a value with little degree of assurance, the values may be the first to question when model results do not match actual outcomes. In such an instance, different values of the parameter can be tried and if they allow the model to reproduce actual outcomes, they should be discussed with the domain experts. This is a case of obtaining pieces for a model, putting them together in the model, and seeing whether they "fit," such that the model can reproduce an actual outcome. If the model output does not fit actual outcome data, then the pieces must be reconsidered, starting with the most suspect or weakest piece.

5.4 ESTIMATING THE PARAMETERS OF A DATA DISTRIBUTION

5.4.1 USING ELICITED DATA

If either a uniform or triangular distribution is indicated, the elicited values specify the distribution. However, if a tailed distribution is indicated, a tool for plotting probability distribution functions can be used to produce a plot that incorporates the elicited information. An example of such a tool is StatAssist by MathWave [19], which is used in this chapter. The following three examples illustrate the use of elicited information.

1. Suppose the following values are elicited for an input parameter: minimum of 40 and maximum of 120. Figure 5.1 is a uniform distribution that fits this description.
2. Suppose the elicited values include a most likely value: minimum 40, most likely 60, and maximum 120. Figure 5.2 is a triangular distribution that fits the description.
3. Suppose the elicited information includes information about a right tail: minimum 40, most likely 60, a right tail with no more than 5% of values higher than 120. In this case a Weibull distribution is chosen, and with StatAssist, parameters are tried until they produce a distribution that fits the elicited description. The result in Figure 5.3 is a Weibull distribution with a shape parameter (α) of 1.5, a scale parameter (β) of 40, and a location parameter (γ) of 40.

Many distributions are available for producing one that fits an elicited description, but a Weibull distribution is often a good starting choice for producing a continuous distribution due to its versatility and its ease of implementation. Most simulation programs support the Weibull distribution. If the modeler needs to provide the im-

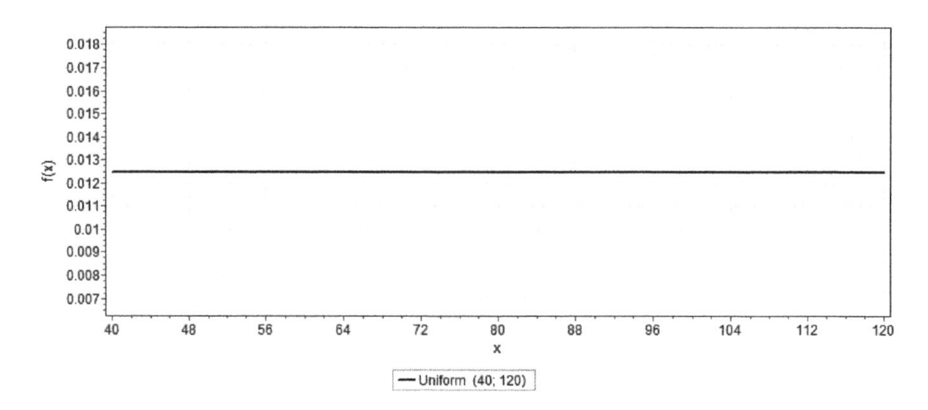

FIGURE 5.1: Uniform Distribution with Elicited Data

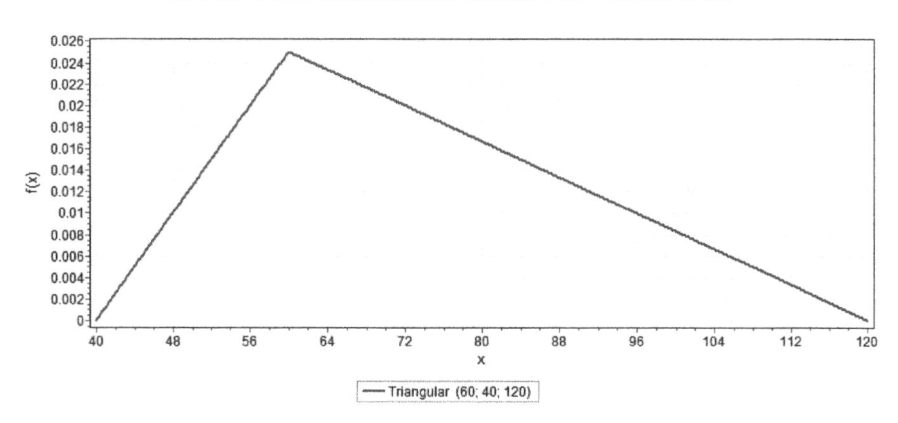

FIGURE 5.2: Triangular Distribution with Elicited Data

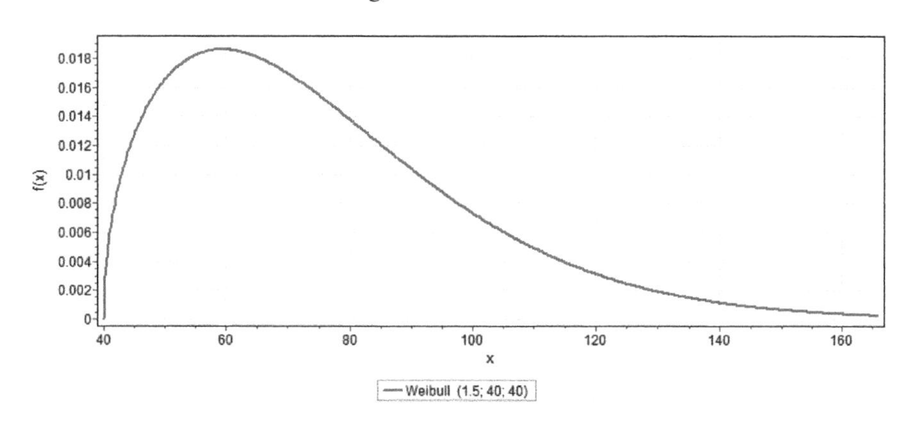

FIGURE 5.3: Weibull Distribution with Elicited Data

plementation, the Weibull distribution has a closed form inverse transform that can be used to generate random variates: $X = \beta(-\ln U)^{1/\alpha}$ where U is a random value from a uniform distribution between 0 and 1, and X is the desired random value from a Weibull distribution.

Producing distributions in this manner is a trial-and-error approach and can be time-consuming, so the modeler is cautioned to ensure that the elicited data is firm enough to warrant the effort. Also apropos is the earlier advice of expending limited effort on distributions until a sensitivity analysis shows that parameters warrant the effort of more precise fitting. Triangular distributions may suffice until then.

5.4.2 USING COLLECTED DATA

Turning from elicited data to data collected through measurement and recorded in a data repository, one should ask about its applicability to a model parameter. The data may not have been collected for the model parameter, but for another purpose. In that case, one must ask about the measurement instruments, recording, and data. To what degree was the data collected subjectively as opposed to use of an objective measurement instrument? Did the instrument measure data that matches the meaning of the model parameter? Was the data measured with sufficient accuracy and recorded with sufficient precision? Was the sample size or frequency of collection sufficient? Is the data complete?

Assuming that the answers to these questions are satisfactory for the modeling purpose, the modeler must examine the data and decide whether to use the data as-is or fit a distribution to it. Sometimes a stream of data is representative of all values encountered during the period of a run. For example, suppose one year of data is available for a set of regularly scheduled tasks and each model run represents one year of processing. In this case, a modeler can choose to read the data stream directly from a table, although a large number of database table reads can slow a model run. Except for cases like the one just mentioned, fitting a distribution to data and sampling the distribution is generally recommended over reading a data stream because a modeler has no concerns for the amount of data needed for a run or for many runs when using a random distribution.

An analysis of data should begin with ascertaining whether the data represents continuous or discrete values. If the data consists of discrete values, then a discrete distribution should be generated from the data, otherwise a continuous distribution will be sought.

The next step in analyzing data is producing a histogram to look at how the data is distributed. One should not assume that data is normally distributed or distributed according to any other particular distribution. Looking at a histogram of data is the best way to form a first impression of the kind of distribution a set of data has.

Choosing the number of histogram bins is automatic in statistical software, but trying different numbers of bins is usually helpful. Given sufficient data, a heuristic rule for the number of data bins is 20 to 30. However, this number of bins may be too large, producing gaps in the histogram. Another heuristic rule is to use a number of bins that produces a histogram without gaps. If a histogram has large gaps or multiple modes, then an empirical distribution may be necessary.

Finally, distribution fitting assumes data independence, so one should examine the data for evidence of autocorrelation. A number of techniques are available for this purpose [25]. The simplest method is to plot sequential values in a scatterplot such that each adjacent pair of values is an ordered pair in the plot for values $x_1, x_2, , x_n$, plot (x_i, x_{i+1}). Independent data will be evenly scattered over an area of the scatterplot as in Figure 5.4, but autocorrelated data will form a pattern such as the example in Figure 5.5. If data is autocorrelated, then uncorrelated samples should be taken or the data should be batched to produce independent mean values. See Chapter 6 for more discussion of these techniques.

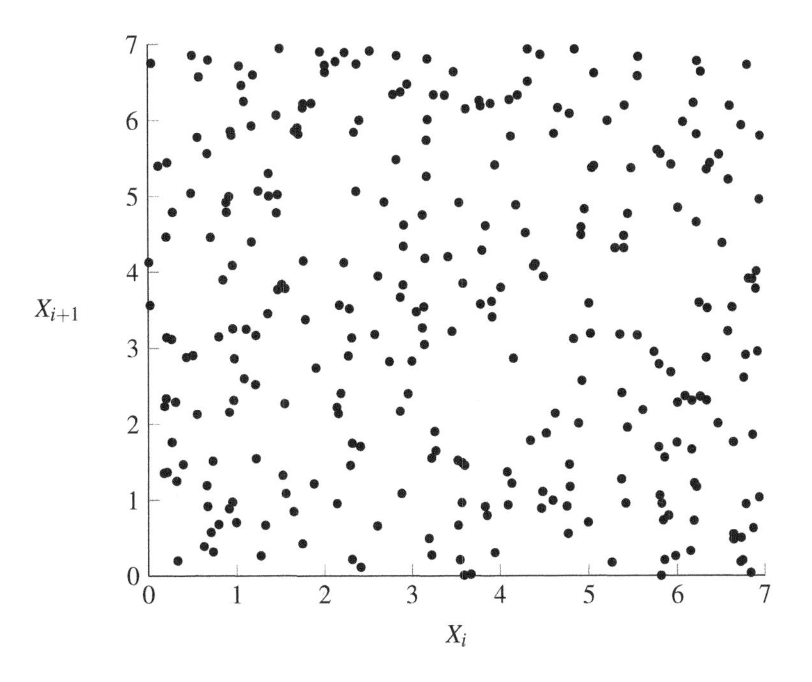

FIGURE 5.4: Sequential Scatterplot of Independent Data

5.4.3 FITTING KNOWN DISTRIBUTIONS TO COLLECTED DATA

Distribution-fitting software is available for performing the calculations necessary to select a distribution that best fits a set of data. These software packages work by trying to fit known distributions to data. For each known distribution, the software calculates estimates of the distribution parameters. For example, data is fit to a normal distribution by estimating values of the mean and standard deviation for which the data most closely approximates a normal distribution. The two most commonly

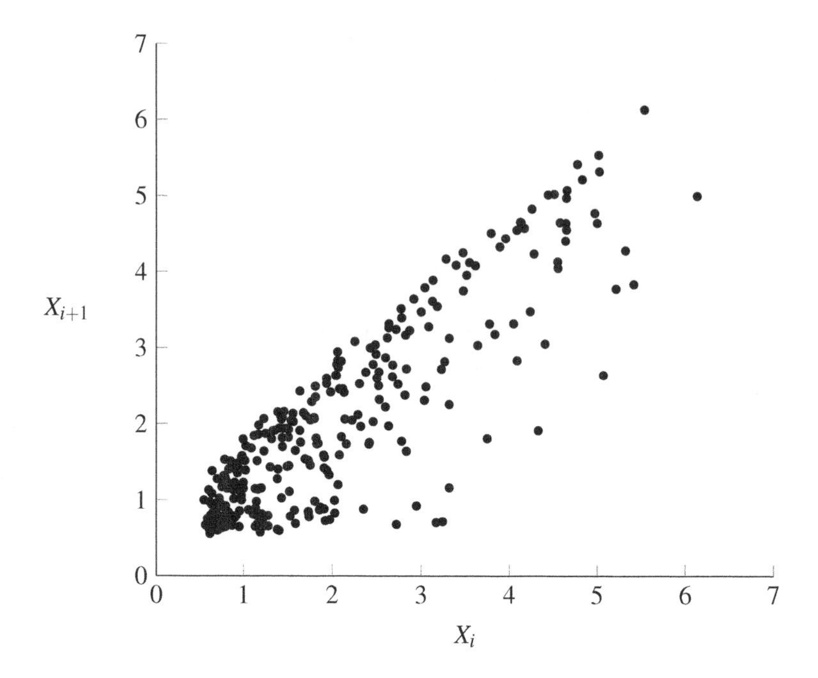

FIGURE 5.5: Scatterplot of Autocorrelated Time in Process Data

used methods for estimating the parameter values of distributions are the maximum likelihood method and the method of moments. The method used by a distribution-fitting software program may depend on the distribution being fitted.

As a distribution-fitting program tries each distribution in its list and calculates parameters for the candidate distribution, it calculates how well each candidate distribution actually fits the data. These calculations are a *goodness-of-fit* test, which is a statistical hypothesis test for which the null hypothesis is that the data belongs to the proposed distribution.

Commonly applied goodness-of-fit tests include the chi-square (χ^2), Kolmogorov-Smirnov (K-S), and Anderson-Darling (A-D). The distribution-fitting software provides a statistic for each of these goodness-of-fit tests and the test statistic is used to rank the candidate distributions. Figure 5.6 is a histogram of example data labeled Ao and Figure 5.7 is a goodness-of-fit summary of the top eleven candidate distributions using StatAssist. Notice that the three goodness-of-fit tests produce statistics that provide different rankings to the distributions. The distributions are sorted by the A-D ranks in Figure 5.6 but can be sorted by either of the other two test rankings.

The χ^2 test measures whether the observed data is significantly different from the proposed distribution. This test is based on dividing the data into intervals and taking the difference between the number of occurrences observed in the data and the number of occurrences expected by the proposed distribution. The test statistic

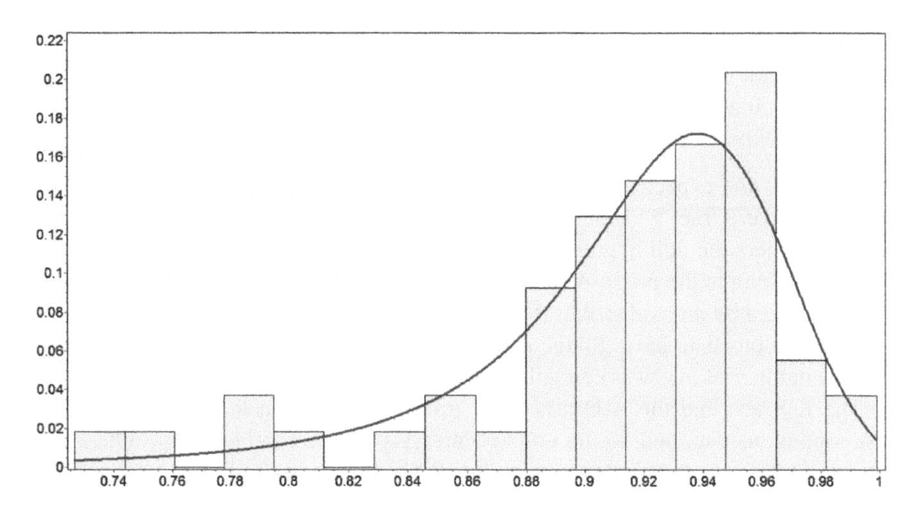

FIGURE 5.6: Histogram of Ao Data with Logistic Distribution Fit

Goodness of Fit - Summary

#	Distribution	Kolmogorov Smirnov		Anderson Darling		Chi-Squared	
		Statistic	Rank	Statistic	Rank	Statistic	Rank
16	Gen. Logistic	0.09075	1	0.39489	1	0.53145	1
43	Weibull (3P)	0.10093	3	0.6057	2	5.0796	6
1	Beta	0.12936	8	0.79538	3	11.541	26
22	Log-Logistic (3P)	0.09811	2	1.3456	4	5.6208	7
34	Pert	0.16233	11	1.3532	5	3.0783	4
2	Cauchy	0.11876	6	1.4895	6	3.2915	5
42	Weibull	0.16198	10	1.5156	7	13.339	28
5	Error	0.1602	9	1.9398	8	3.0273	3
24	Logistic	0.16742	12	1.9649	9	8.9232	12
33	Pearson 6 (4P)	0.18086	13	2.2273	10	8.7535	9
27	Normal	0.18363	16	2.2595	11	8.8259	11

FIGURE 5.7: Goodness of Fit Summary for Ao Data

χ^2 is calculated for the χ^2 test according to Equation 5.1.

$$\chi^2 = \sum_{j=1}^{k} \frac{(N_j - np_j)^2}{np_j} \tag{5.1}$$

where

n is the total number of observations

N_j is the number of X_i's in the jth interval

p_j is the expected proportion of of X_i's that would fall in the jth interval

Since np_j is the expected number of X_i's that would fall in the jth interval if the hypothesis were true, χ^2 is small if the fit is good. Therefore a large value of χ^2 is reason to reject the null hypothesis and accept the alternate hypothesis that the data does not belong to the proposed distribution. Because the test uses discrete intervals and is affected by interval size, it is better suited to discrete data. However, it is used with continuous data having large numbers of data points, subject to the limitation that the number of intervals can influence the test statistic.

The K-S test and the A-D test differ from the χ^2 test in that they are used to test continuous data and do not employ intervals for the sample data. Rather, they compare two cumulative distribution functions, the proposed distribution, and the empirical cumulative distribution of the data. The K-S test measures the largest discrepancy in their cumulative frequencies for each value in the sample data. Because the test statistic is the largest discrepancy between two cumulative distribution functions, a smaller value indicates a better fit. Also, it is a stronger test than the χ^2 test.

The A-D test also compares two cumulative distribution functions but weights the differences in the tails of the distributions more heavily because differences are more likely to occur in the tails. Thus, the A-D test is more stringent than the K-S test, which treats all differences equally.

A large value of a test statistic indicates a poor fit. For the K-S and A-D tests, critical values that vary by distribution are used to accept or reject the hypothesis that the data fits each distribution. Good distribution-fitting software uses these critical values to reject a distribution from a list of candidates.

Probability plots provide graphical means of judging adequacy of data fit to a distribution. The Quantile-Quantile (Q-Q) plot compares data of two distributions, the proposed distribution and the data's empirical distribution by quantiles (also called percentiles). For each percentile of the proposed distribution, the data for the same percentile is plotted. Figure 5.8 provides an example of a Q-Q plot for the Ao data and the General Logistic distribution with StatAssist.

The Probability-Probability (P-P) plot contains the two cumulative distribution functions, the proposed distribution, and the data's empirical distribution. Figure 5.9 provides an example of a P-P plot for the Ao data and the General Logistic distribution.

The difference between a Q-Q plot and a P-P plot is similar to the difference between the χ^2 test and the K-S and A-D tests. The former considers data noncumulatively and the latter considers data cumulatively. The Q-Q plot shows differences between the data and the proposed distribution in the tails more clearly and the P-P plot shows differences between the data and the proposed distribution around the mode more clearly. Both the P-P and the Q-Q plots are usually available in distribution-fitting software.

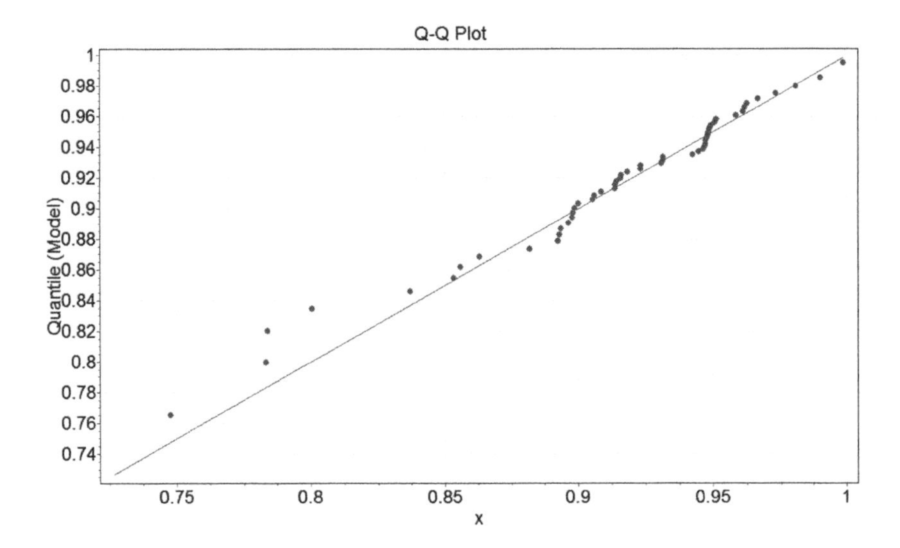

FIGURE 5.8: Q-Q Plot for the Ao Data and the General Logistic Distribution

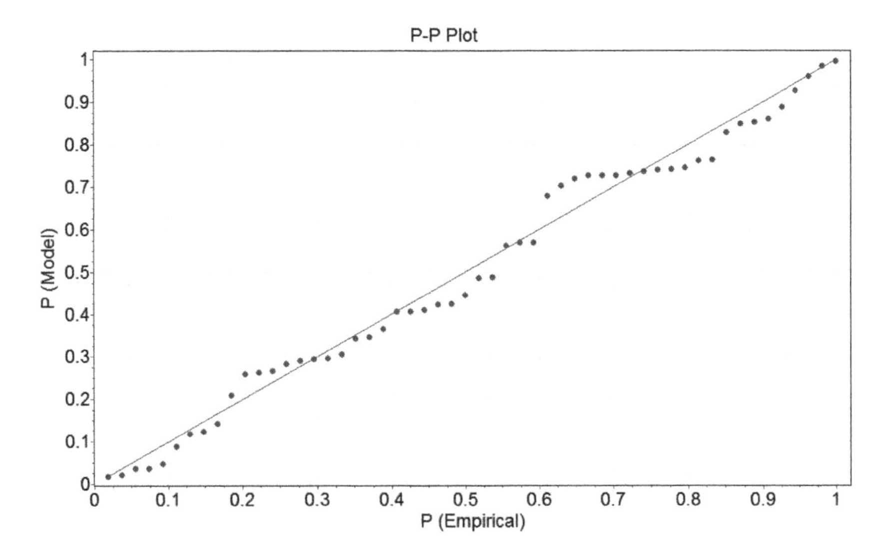

FIGURE 5.9: P-P Plot for the Ao Data and the General Logistic Distribution

Selecting a distribution from the list of candidates may not be a matter only of choosing the best fit. The simulation software one uses must also support the distribution. Most of the distributions shown in Figure 5.7 are typically supported, but suppose a distribution that provides the best fit is not supported by one's simulation software. In that case, one can usually choose the next best distribution.

Example: Goodness-of-Fit Test

For a manual example of goodness-of-fit testing, we will apply the Kolmogorov-Smirnov test to collected data. The following five steps are performed for the Kolmogorov-Smirnov test:

1. Rank the field data in increasing order.
2. Compute the following using the theoretical CDF:

$$D+ = max[i/NF(x_i)] \qquad 1 \leq i \leq N$$
$$D- = max[F(x_i)(i-1)/N] \qquad 1 \leq i \leq N$$

3. Let $D = max(D+, D-)$.
4. Find the critical value from a Kolmogorov-Smirnov table for a given significance level and sample size.
5. If D ¡= the critical value, accept the candidate distribution as having a good fit to the field data. Reject it otherwise.

We want to assess whether a set of data on service times is uniformly distributed between 1 and 2 with a significance level of $\alpha = .05$. Suppose we have five observations on the time in minutes as follows: 1.38, 1.25, 1.06, 2.0, and 1.56. For the uniform distribution the CDF is:

$$F(x_i) = x/(b-a) \qquad a \leq x \leq b.$$

When $b = 2$ and $a = 1$, $F(x_i) = x/2$. Table 5.1 shows the calculations.

TABLE 5.1
Kolmogorov-Smirnov Calculations

i	x_i	$F(x_i)$	i/n	$i/n - F(x_i)$	$F(x_i)(i-1)/n$
1	1.06	.53	.2	-.33	.53
2	1.25	.625	.4	-.225	.425
3	1.38	.69	.6	-.09	.29
4	1.56	.78	.8	.02	.18
5	2.0	1.0	1.0	0	.2
				D+ = .02	D- = .53

The final result shows that $D = max(.02, .53) = .53$. The critical value from Kolmogorov-Smirnov tables for a sample size of 5 and significance level of .05 is .565. D is less than the critical value, so the hypothesis that the distribution of data is uniform between 1 and 2 is not rejected.

5.4.4 CONSTRUCTING EMPIRICAL DISTRIBUTIONS

Sometimes distributions cannot be found that provide a good fit to a dataset, perhaps because an insufficient amount of data is available. Attempts at assuming a distribution can provide very poor results when comparing outcomes with original data. For example, in one case a modeler took the product of three input variables for crew size, task duration, and number of tasks per month to obtain monthly effort. The dataset of each of the three input parameters was small and each dataset exhibited a non-continuous distribution. Assuming triangular distributions for each of the three input variables produced monthly effort approximately three times what was justified by the raw data. Producing a distribution is a generalization of data for the sake of characterizing data and reapplying it. It must be done accurately. In the case of the labor inputs, empirical distributions had to be used.

An empirical distribution is simply a cumulative distribution of the data. For example, suppose twelve discrete values (3, 15, 21, 24, 4, 7, 14, 22, 14, 10, 7, 18) are obtained for the parameter Jobs per Month. Table 5.2 is constructed for this in which the first row, Jobs per Month, contains all the values in the range of the data. The occurrences of these values will be counted for a histogram. (Note that constructing the empirical distribution does not require all discrete values in the range, specifically the values not in the data are not required. They are included here for the sake of illustration and for ease of producing the histogram and distribution plots.)

TABLE 5.2
Empirical Distribution of Jobs per Month

Jobs per Month	0	1	2	3	4	5	6	7	8	9	10
Frequency	0	0	0	1	1	0	0	2	0	0	1
Cum. Frequency	0	0	0	1	2	2	2	4	4	4	5
Cum. Probability	0.00	0.00	0.00	0.08	0.17	0.17	0.17	0.33	0.33	0.33	0.42

Jobs per Month	11	12	13	14	15	16	17	18	19	20	21
Frequency	0	0	1	2	0	0	0	1	0	0	1
Cum. Frequency	5	5	6	8	8	8	8	9	9	9	10
Cum. Probability	0.42	0.42	0.50	0.67	0.67	0.67	0.67	0.75	0.75	0.75	0.83

Jobs per Month	22	23	24
Frequency	1	0	1
Cum. Frequency	11	11	12
Cum. Probability	0.92	0.92	1.00

The Frequency rows in Table 5.2 contain the number of occurrences of each value. Figure 5.10 is the histogram produced from the Frequency rows. This histogram clearly indicates that the data does not fit a known distribution. Therefore, an empirical distribution is used. To produce the empirical cumulative distribution, the Cum.

Frequency rows accumulate the values in the Frequency row. The empirical distribution is in the Cum. Probability rows where the Cum. Frequency is divided by the total number of Jobs per Month. This distribution can be specified as the ordered pairs:

[3, 0.08), [4, 0.17), [7, 0.33), [10,0.42), [13, 0.50), [14, 0.67), [18, 0.75), [21, 0.83), [22, 0.92), [24, 1.00)

where the open square brackets indicate "greater than or equal to," the closed parentheses indicate "less than," and the closed square bracket indicates "less than or equal to." When plotted, an empirical distribution of this discrete data appears as a step function in Figure 5.11.

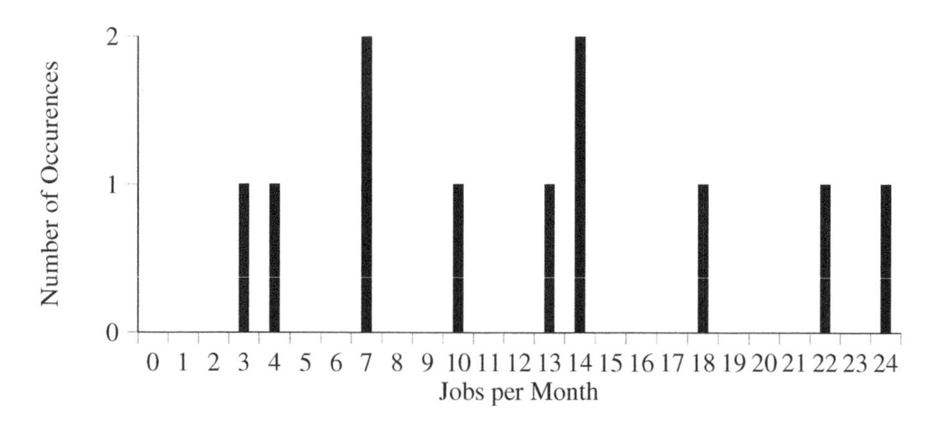

FIGURE 5.10: Histogram of Jobs per Month

Implementing an empirical distribution of discrete data as in this example is straightforward. A random value between 0 and 1 is generated, then used to search for a value greater than or equal to one in the cumulative probability row and less than the next value. The returned value is the next value in the data row. For example, suppose a pseudorandomly generated value of 0.516 is used to search Table 5.2. The random variate is greater than or equal to 0.50 and less than 0.67, therefore a value of 14 is returned. In terms of Figure 5.11, 0.516 intersects the vertical segment over the x value of 14, so 14 is returned.

The preceding example used discrete data. For continuous data, constructing an empirical distribution is similar. For example, twelve continuous values (8.9, 4.6, 1.6, 3.5, 11.8, 1.5, 6.0, 7.0, 1.8, 1.9, 2.8, 1.9) are obtained for the parameter Job Duration. Table 5.3 is constructed similar to Table 5.2, but using only the data values (which would ordinarily be done for discrete data as well). Figure 5.12 is a plot of this distribution.

Because the data is continuous, Figure 5.12 does not have the appearance of a

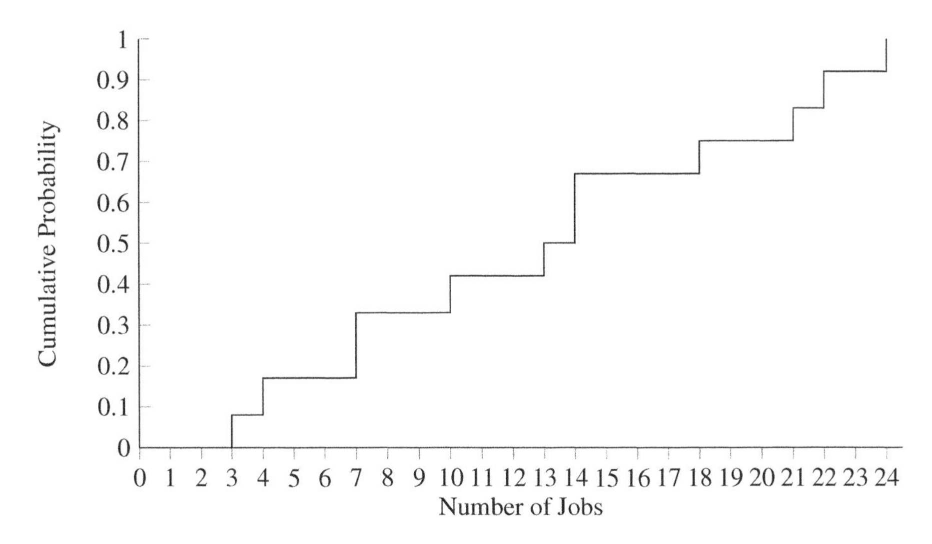

FIGURE 5.11: Empirical Cumulative Distribution of Jobs per Month

TABLE 5.3
Empirical Distribution of Job Durations

Job Duration	1.5	1.6	1.8	1.9	2.8	3.5	4.6	6.0	7.0	8.9	11.8
Frequency	1	1	1	2	1	1	1	1	1	1	1
Cum. Frequency	1	2	3	5	6	7	8	9	10	11	12
Cum. Probability	0.00	0.09	0.18	0.36	0.45	0.55	0.64	0.73	0.82	0.91	1.00

step function between discrete values. Also, the function for returning a value from the distribution interpolates sample data values as in the following steps.

1. Generate a pseudorandom value x between 0 and 1.
2. Using the table for the empirical distribution, find the largest cumulative probability, $p(i)$, less than or equal to x and its corresponding data item, $X(i)$, in the empirical data. The position of this point in the data sequence is i.
3. Interpolate the variable value: $X_i + (X_{i+1} - X_i)(x - p_i)/(p_{i+1} - p_i)$

As an example, suppose 0.43 is a randomly generated value for obtaining a value x_n of Job Duration:

$$x = 0.43,\ p_i = 0.36,\ p_{i+1} = 0.45,\ X_i = 1.9,\ X_{i+1} = 2.8$$
$$i = 4$$
$$x_n = 1.9 + (2.8 - 1.9)(0.43 - 0.36)/(0.45 - 0.36) = 2.6$$

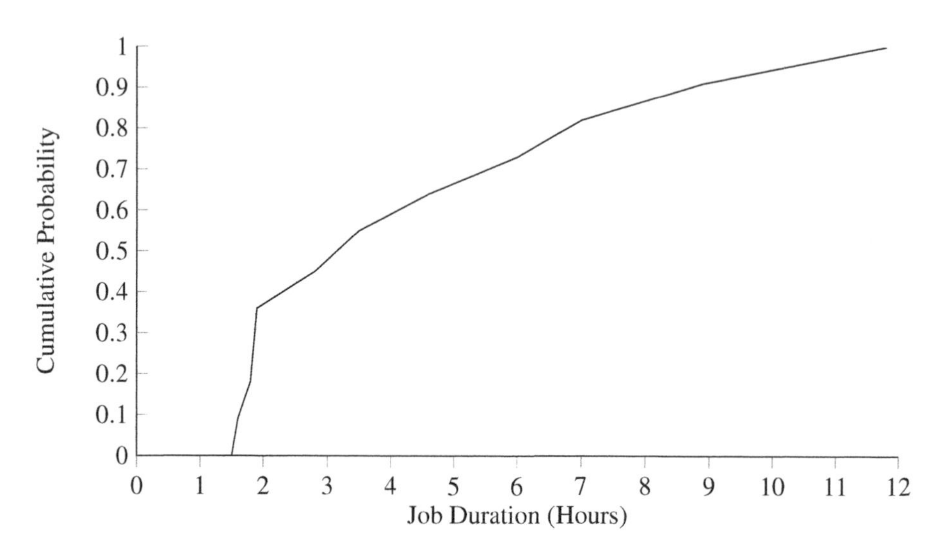

FIGURE 5.12: Empirical Cumulative Distribution of Job Duration

5.5 SUMMARY

The quality of a simulation model depends on the structure of the model and on the quality of the model inputs. Every model input should be considered for its uncertainty. This chapter has discussed how to ensure the quality of input parameter values and has shown applied examples.

Steps performed for input analysis are as follows:

- Identify input variables.
- Decide on the types of input values.
- Collect and/or elicit data on the input variables.
- Estimate the parameters of a data distribution.
- Fit known distributions to data, test fits, and select a distribution.

Input parameters have to be chosen such that they are readily recognizable to stakeholders, they define the model boundary according the model purpose and scope, and they respect the level of abstraction required of the model. Sources of data must also be considered when input parameters are chosen.

When deciding on types of input variables, one can choose between constants and variables for parameters. Constant values may be used when one knows an input is actually constant, when one wants to start modeling with a value until variable data are available, or when running designed experiments with selected values. Otherwise, input parameters are considered variables that take on randomly distributed values.

Collecting data for constructing distributions can be challenging and time-consuming. If a modeling purpose allows for use of elicited information, then such information can be used to produce distributions. Eliciting data from domain experts is necessary when no data exists for a model parameter and the cost or duration of data collection is prohibitive.

When estimating the parameters of a data distribution and recorded data are available, it should be analyzed with distribution-fitting software. Goodness-of-fit tests provide rankings of candidate distributions to assist in choosing a best fitting distribution for data. Some common tests include chi-square (χ^2), Kolmogorov-Smirnov (K-S), and Anderson-Darling (A-D). Distribution-fitting software provides test statistics for these goodness-of-fit tests to rank the candidate distributions. If a good fit cannot be found, then an empirical distribution can be used.

When information about skewedness is provided, plotting distributions can be done in a trial-and-error manner until one finds a distribution that fits the description of the elicited information. Goodness-of-fit can also be assessed graphically using probability plots for the P-P plot and the Q-Q plot.

6 Simulation Runs and Output Analysis

6.1 INTRODUCTION

The title of this chapter might lead one to think that output analysis starts after a model has been constructed and verified and is being run to produce results of interest to stakeholders. Actually, output analysis can be considered as beginning with engagement (Section 2.2) and model specification (Section 2.3). The first step in output analysis is choosing output parameters that 1) are relevant to stakeholders' concerns and 2) provide data for a modeler to analyze for producing information for stakeholders. Therefore, output parameters must serve the modeling purpose and scope directly. Sometimes the first thing a modeling sponsor wants to know when considering a modeling project is what information they will see. This query provides the modeler an opportunity to discuss modeling benefits in terms of outputs that interest the sponsor and the variety of analyses that can be employed to produce information useful to the sponsor. As an engagement is undertaken and a model specified, the modeler can return to the sponsor with a plan for producing information from specified output parameters and analysis techniques.

Having begun the modeling process with the end in mind, a modeler can continue through the process with the goal of producing the desired outputs. The modeler will be mindful of what is necessary to produce values adequate for the output parameters during the modeling paradigm selection (Section 2.4.1), construction (Section 2.4.2), data collection and input analysis (Section 2.4.3 and Chapter 4), and assessment (Section 2.5). The advice of focusing a model on a question is particularly helpful because output parameters and analysis can be chosen to address the question.

Consider that systems and processes are rarely, if ever, deterministic. Most systems are stochastic, exhibiting randomly varying output values regardless of the constancy of input values. This feature of a system not reproducing the same output values with the same inputs is represented in random variables for stochastic models. On the other hand, deterministic modeling may be chosen at a level of abstraction that ignores this variation in model parameters. Even so, one can systematically vary input values and measure variation in output values with deterministic models.

With either approach, stochastic or deterministic, variation in outputs (whether due to randomness, input manipulation, or both) is the basis for statistical analysis of outputs. This chapter addresses statistical analysis of outputs for stochastic models, with the understanding that some of the analytical techniques can also be applied to deterministic models by varying input values systematically.

6.2 PRELIMINARY CONSIDERATIONS

Before discussing statistical analysis of output values, several points need to be introduced. First, simulation is a statistical sampling experiment in which models convert stochastic inputs into statistical data output. Because simulation is a sampling process, variable estimates are subject to sampling choice and sample size. Since random samples from probability distributions are used as simulation inputs, the outputs (estimates) for each run are themselves random variables that may have large variances.

The following example shows how results may vary widely across a small number of runs. The electric car charging station model from Chapter 3 was executed using exponential interarrival and service time input distributions. It was run 10 times simulating a 12-hour time frame of system operations. Note in Table 6.1 how widely the primary outputs vary across the 10 replications. It would be risky to draw conclusions from any single run about the true system parameters.

TABLE 6.1
Results of 10 Charging Station Model Runs

Run Number	Cars Served	Mean Waiting Time (Min.)	Resource Utilization	Average Queue Length
1	118	11.3	0.8	1.9
2	115	3.8	0.7	0.7
3	117	8.6	0.7	1.4
4	111	8.5	0.8	1.4
5	105	8.9	0.7	1.4
6	107	15.5	0.8	2.3
7	107	19.8	0.8	3.1
8	99	9.7	0.7	1.3
9	125	24.7	1.0	4.3
10	114	12.2	0.7	2.0

A single run, or even a small number of runs, is insufficient if the M&S goal is measuring certain output variables for a given set of inputs. Conclusions from model runs should be based on a sufficient number of runs with sufficient variation of input values. If one is comparing variable characteristics under different conditions, then many runs must be made for each condition. If one is comparing system alternatives to improve a system or design a future one, then multiple runs are warranted for each alternative configuration. Choosing a number of runs will be discussed further in this chapter.

The next point is that most real processes are *nonstationary*, that is distribution of their measured outputs varies over time. Data collected at one time may have a different distribution from data collected at another time. Thus, validation of a model that tests model data for the same distribution as data collected for the real system may not be possible. Consequently, a modeler should be concerned with whether differences in real and modeled output are significant. These differences can be measured using confidence intervals on differences in mean values and checking whether the intervals include 0 with methods such as those described later in Section 6.5.2.

Also of concern is *autocorrelation* in output data, both real and modeled. Adjacent entities moving through a system or model are subject to delays and the processing of leading entities influence the processing of following entities. Therefore, output measures that reflect such processing are correlated. Statistical techniques ordinarily assume that data values are independent, so care must be taken in how data is prepared for statistical analysis. One method for addressing autocorrelated data relies on the Monte Carlo technique described in Chapter 4: use different random number streams across runs, producing an independent replication in each run.

At this point, the number of replications becomes a consideration. How many Monte Carlo replications are enough to approximate properties of a true sampling distribution? Each data set yields a draw from the true sampling distribution, so standard statistical techniques are used to select a sample size that will achieve acceptable precision of the approximation of output values. Methods for obtaining independent samples are discussed further in subsequent sections of this chapter.

The next major point to be introduced before proceeding with discussion of statistical analysis is that of differentiating simulation types and system states in the next section.

6.3 SIMULATION TYPES FOR OUTPUT ANALYSIS

It is important to distinguish between system classes and states for performing output analysis. The analysis method may differ depending on whether the system is a *terminating system* or *non-terminating system*; and whether the system is in a *transient state* or *steady state*. Terminating systems usually start from no-action or an empty state and end with either of the same. Termination occurs after a time delta or event occurrence. Examples of time lapse terminations include daily service operating hours or quarterly manufacturing operations. Examples of event-based termination include device failure or project completion. Terminating systems may or may not reach steady state. If a steady state exists, the system may be treated as a non-terminating for some analysis purposes.

Non-terminating systems have no end to operations for a practical time horizon. Examples of continuously operating systems include computer and phone networks, emergency rooms, road traffic, and many factories when the daily starting time is same as the of end of the previous day. A non-terminating system usually, but not always, has a steady state. One may wish to study the transient and/or steady state conditions of a system. A non-terminating system can also be treated as a terminating system to study transient period behavior.

An example demonstration for considering system states due to transient effects is shown in Figure 6.1. It shows the results of four runs for the electric car charging scenario whereby the cumulative mean waiting time is updated at each event indicated by the circle markers. Lines of demarcation are shown at the transition times between transient and steady states. Using statistics in the transient state (its range also varies across runs as illustrated) would not reflect the vast majority of operations and events. Conclusions drawn would be misleading. Using too short of a run length would also not capture the essential behavior. In this case the transient state could be lopped off and only the steady-state portion analyzed. Conversely the transient state may be a focus of concern, in order to determine the startup period for better system insight and operational planning. It is thus important to consider the time period(s) used in a simulation armed with knowledge of the system state behavior.

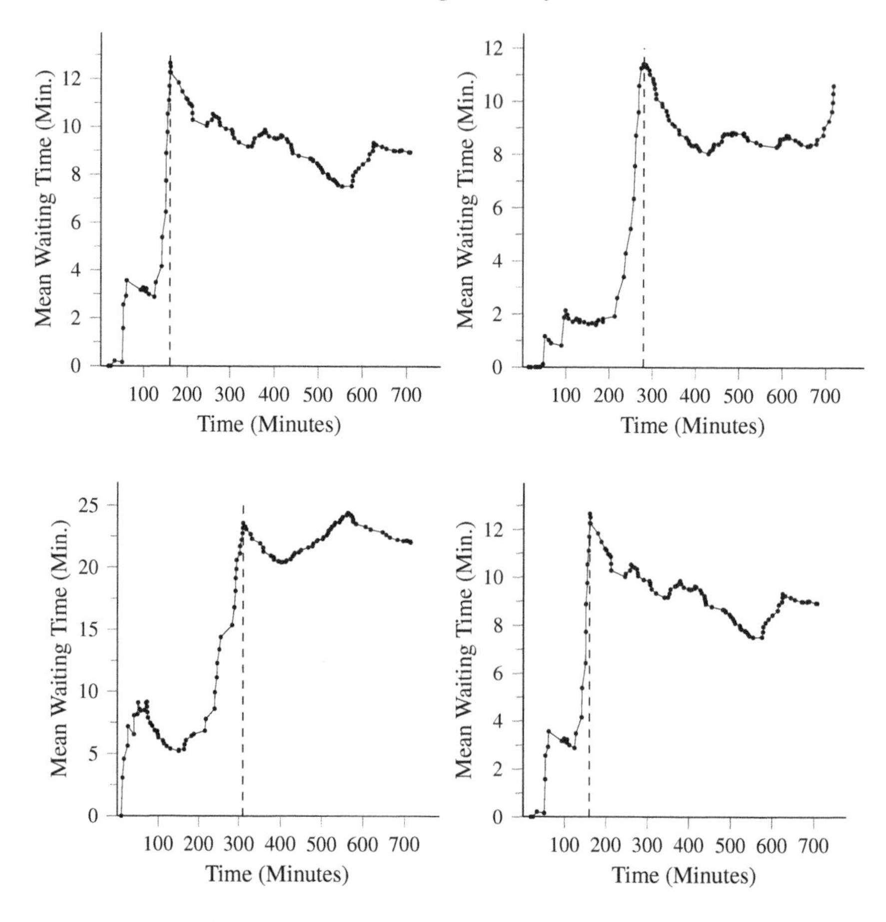

FIGURE 6.1: Transient and Steady States for Mean Waiting Time

6.4 MODEL EXECUTION

6.4.1 SETTING UP RUNS AND DESIGNING EXPERIMENTS

In experimental design we decide what configurations to run before simulation. Sets of input values are designed for analysis that yields information about the relationships between input parameters and output parameters. The sets of values are selected to obtain maximum information from a minimum number of runs. Experiments are usually run iteratively with each experiment yielding more insight into system behavior.

Choosing what kind of runs to make and how many of each requires careful consideration of several related issues. These include the overall consideration of achieving independent samples, run length, transient effects and steady-state periods, the number of runs, and settings of input variables for controlled experimentation. Recall from Chapter 4 that when using the Monte Carlo approach more runs produced a higher-fidelity output distribution. Accordingly, more runs produce tighter confidence intervals for the output analysis methods described in Section 6.5.

Each run produces a statistical sample, and determining a proper sample size may require some knowledge of the parameters to estimate. Statistical variance normally decreases with increased sample size, which requires more iterations. The formulas for confidence intervals in Section 6.5 can be manipulated to determine the required sample size to attain a desired confidence interval width in conjunction with an estimated sample variance.

Run-length decisions must consider transient effects over time so that steady state measures are used for analysis and comparative studies. For example, statistics derived from beginning time segments of a run may be invalid because a steady state is not yet achieved or cumulative monotonic trends are not accounted for. A system often starts from an idle state and queues build up to average (steady-state) conditions after the initial customers or events. The electric car charging model example exhibits transient effects that may be masked with an improper run length and no resulting steady state.

A variety of different *scenarios* are run depending on the modeling goals and questions. A parameterized model is known as a scenario when the inputs are set or calibrated for specific system conditions. A comparison of system alternatives involves different scenarios being run with parameter settings reflecting the various options.

An experiment in our context is the execution of a computer simulation model. Experimental design is a way of deciding before runs are made which configurations to simulate so that the desired information can be obtained with a minimal number of runs. Carefully designed experiments will eliminate the hit-and-miss experience. Experimentation should lead to an understanding of the simulated system and how to improve it.

In some cases, trial and error are necessary to try a large number of parameter values for optimizing system performance. Finding the optimal values for a system may require a heuristic process. Otherwise, trying all possible combinations of variables can be prohibitively time-consuming. Note that some simulation packages enable some degree of automated optimization.

When designing simulation experiments, the input parameters are called *factors*, and the output performance measures are called *responses*. Which factors to keep fixed and which should be varied depends on the goals of the study. Factors are also classified as controllable or uncontrollable depending on whether or not they represent actual options in the real-world system. Normally simulation focuses on the controllable factors.

The experimental design for a one-factor model is straightforward. Run the simulation at various values, or levels, of the factor. A confidence interval can be formed for the expected response at each of the factor levels. Suppose now there are k factors and we want an estimate of how each factor affects the response, as well as whether the factors interact with each other. The number of runs increases exponentially with the number of factors and each factor must be set at specific levels. A $2k$ factorial design is an economical strategy to measure interactions. It requires only two levels for each factor and then calls for simulation runs at each of the $2k$ possible factor-level combinations.

Example: Effect of Sample Size (# Replications)

To demonstrate the effect of sample size, the charging station model was used in four experiments of 25, 100, 500, and 1000 runs. Each run in each experiment provided an independent replication and a sample data point. Figure 6.2 shows the resulting mean waiting time output histograms (uniform distributions were used for arrival and service rates). Holding all other inputs fixed, it is evident visually that the PDF histogram becomes better defined and filled out as the number of model runs increases. The same results occur when varying any of the other conditions of the simulation. More replications produce a better approximation. For a large number of applications, 1,000 iterations is sufficient, and rarely are more than 100,000 runs warranted.

Example: Effect of Run Length

A scenario with the charging station model will illustrate that determining the proper run length must consider long-term trends. In the scenario the charging station has inadequate service resources, and the average behavior is that the queue keeps building. For a given interarrival rate and service time distribution, a single charger cannot keep up with the demands. From an initial idle state, the average waiting time monotonically increases with the simulation length shown in the Monte Carlo results in Figure 6.3. Deducing waiting time from one of the shorter duration experiments is not representative because the system never achieves steady state.

Steady-state conditions are never attained in this scenario when the simulation is for a 12-hour time frame. The system would be reset during downtime but keep repeating the unacceptable operations when it is operational. If it was a non-terminating 24-hour system with constant traffic it would be unsustainable.

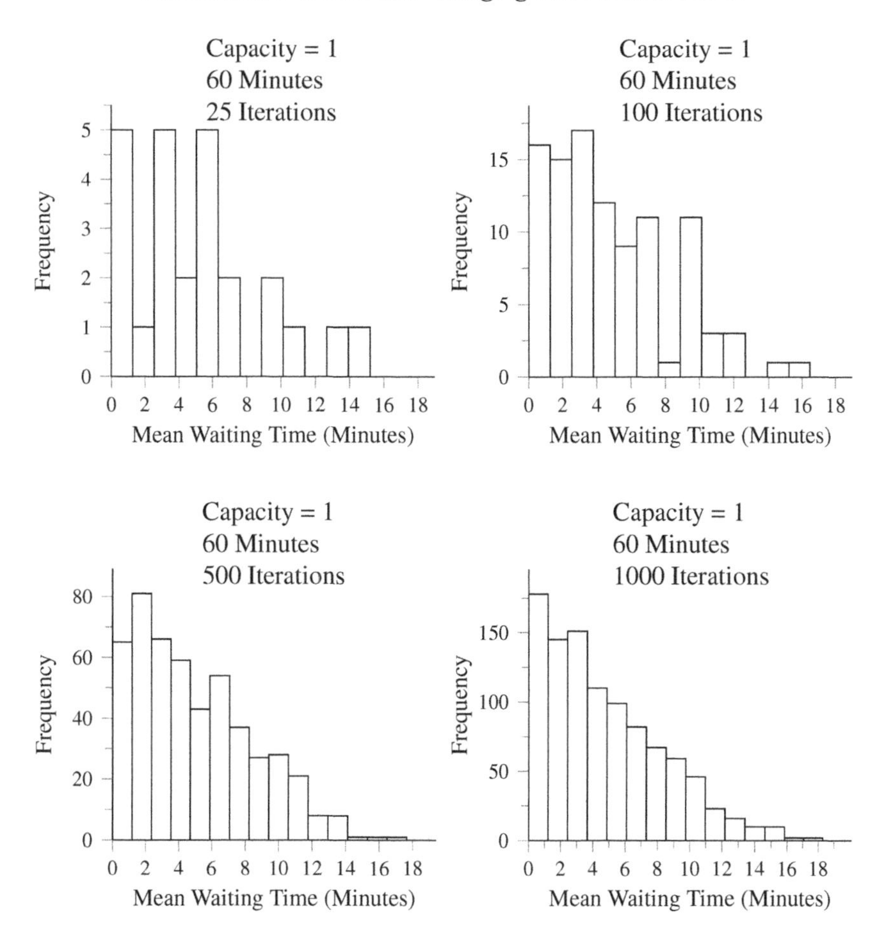

FIGURE 6.2: Effect of Sample Size (# Replications)

Example: *2k* Experimental Design

An example of experimental design will be applied to the charging station scenario. There is a very broad range of options for the number of charging bays and charger types, all across different traffic density profiles specified as arrival rates. It will be simplified with two settings for each of the three variables. Hence the number of factor combinations, or design points, is $2^3 = 8$. The factors for number of charging bays and charger types are controllable and the arrival rate is uncontrollable, i.e., it cannot be affected by the manufacturer and is outside of the decision space.

The chosen values straddle the "sweet spot" of operations whereby most potential

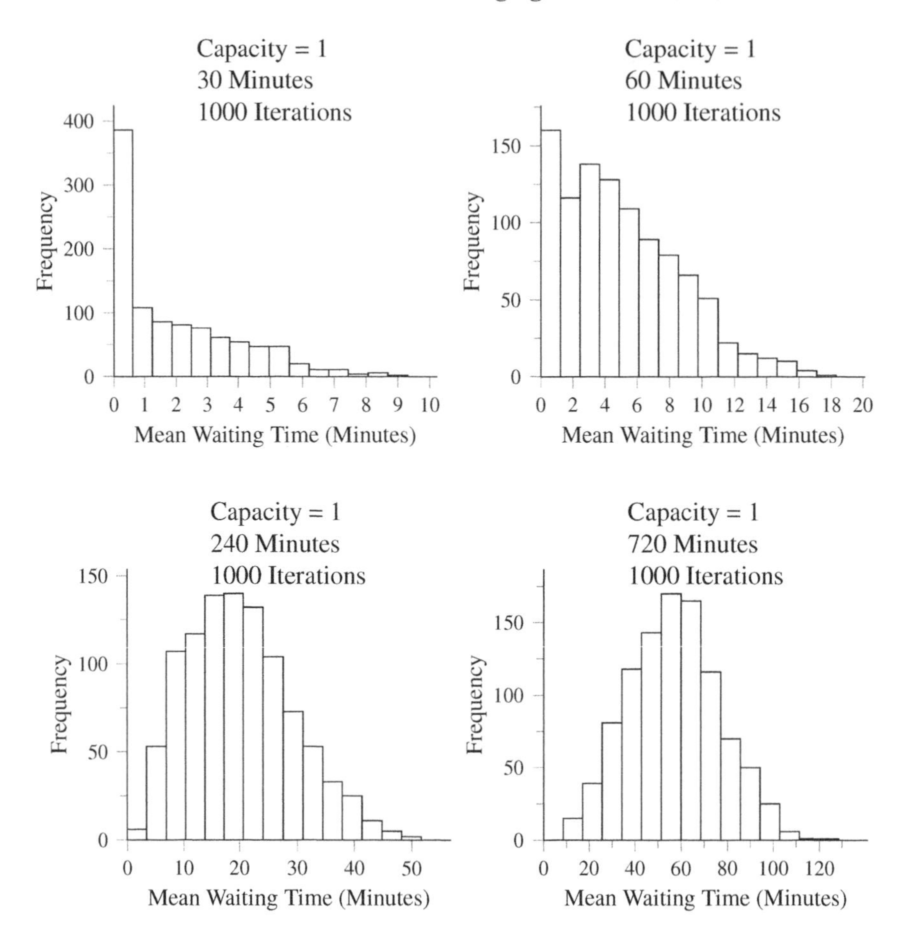

FIGURE 6.3: Effect of Run Length for Non-Steady System

locations are in the 20 to 30 cars-per-hour arrival range. At these nominal operating conditions, the effect of each controlled decision variable combination can be assessed in terms of overall influence in performance measures such as waiting times and utilization percentage. It is assumed that a supercharger is twice as fast as a standard charger in the model. Table 6.2 lists the factor settings and Monte Carlo simulation results (using 1000 iterations) for the Mean Waiting Time.

The results of the experiments are plotted on two types of graphs to understand the effects and interactions of the factors. Figures 6.4 and 6.5 are *main-effect* plots. For each plot, the average waiting time at a particular setting for the factor of interest is the average of the sample means over the two levels of the other factor. In Figure 6.4

TABLE 6.2
***2k* Experimental Design and Results**

Factor Combination	Arrival Rate (Cars / Hour)	Charging Bays	Charger Type	Mean Waiting Time (Min.)
1	20	3	Standard	1.0
2	30	3	Standard	6.1
3	20	4	Standard	.21
4	30	4	Standard	1.0
5	20	3	Supercharge	.06
6	30	3	Supercharge	.20
7	20	4	Supercharge	.01
8	30	4	Supercharge	.03

the values are averages across the charger types. For example, the displayed Mean Waiting Time for 3 charging bays at 30 Cars/Hour is the average for factor combinations 2 (standard) and 6 (supercharge): $(6.1+.2)/2=3.1$. Similarly in Figure 6.5, the values for each charge type are averages across the number of charging bays.

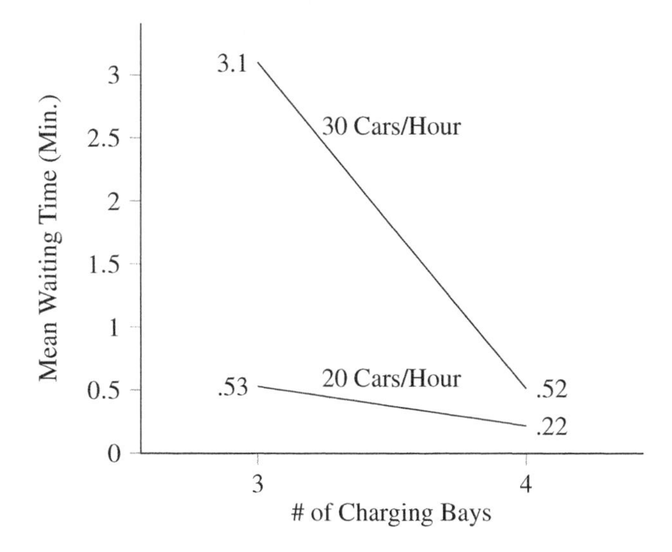

FIGURE 6.4: Main Effect Plot for Charging Bays

The Figures 6.6 and 6.7 are *interaction plots* that show significant interactions indicated by nonparallel lines. In Figure 6.6 for example, it is seen that 3 charging bays increases the waiting time more significantly than 4 charging bays between the

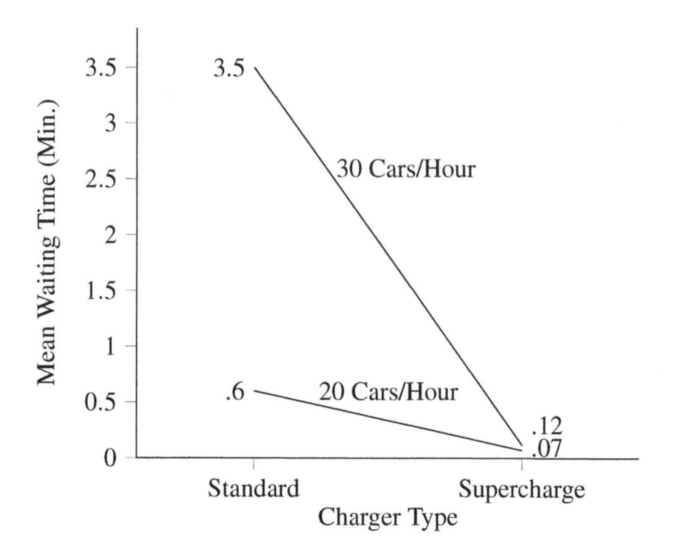

FIGURE 6.5: Main Effect Plot for Charger Type

charger types. Collectively with these plots it is evident that both charger type and the number of chargers have major effects on the waiting time, but the relative changes depend on the other factor and vice versa. The relative differences of effects for the controlled factors also depend on the arrival rate.

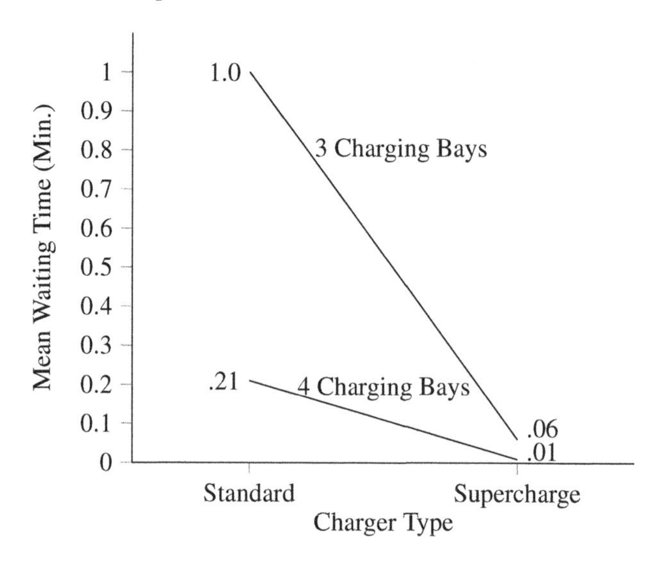

FIGURE 6.6: Interaction Plot of Charging Bays and Charger Type at 20 Cars/Hour

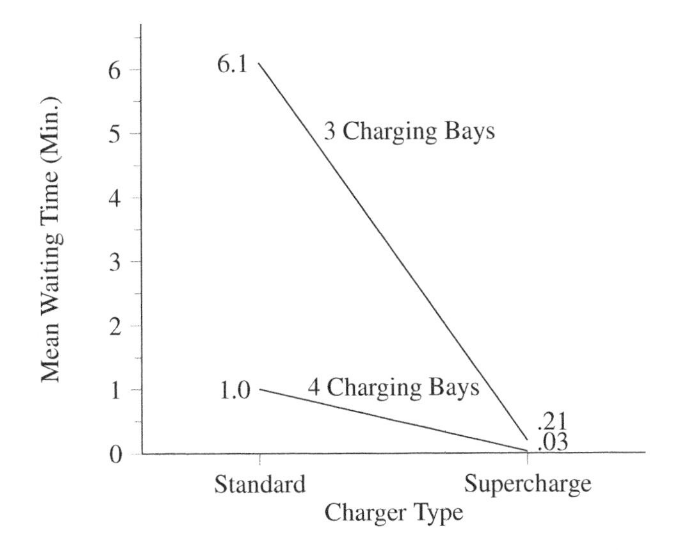

FIGURE 6.7: Interaction Plot of Charging Bays and Charger Type at 30 Cars/Hour

See Section 7.4 for another example of choosing simulation experimental conditions for an extended case study. The reader should also consult advanced statistical texts for more information on experimental design including [21] and [9].

6.4.2 ACHIEVING DATA INDEPENDENCY

Independence of sample data allows use of classical statistical analysis. However, most discrete event simulation models lack data independence because queueing processes usually lead to autocorrelation. Thus the results of the simulation experiments are almost always autocorrelated.

An example of autocorrelated data is the waiting time of transactions in a simple queueing system with a single queue, single server, and exponentially distributed arrival and service times. The behavior of a service transaction will obviously be influenced by the preceding one. If a transaction is delayed for a long time, the following transaction in a queue will also have to wait for a long time. This effect will persist until the queue becomes empty. It is only when all queues are empty that transaction behavior is independent of past history.

Thus simulation experiments must be planned carefully and adequate time spent on the analysis of the results. Two approaches may be taken for autocorrelation. Experiments may be planned so as to eliminate the autocorrelation effects or to use them advantageously. To eliminate autocorrelation effects, one may choose to select samples far apart in the output data. To use autocorrelation advantageously, independent replications or batch means are available.

Replications (single runs) are statistically independent repetitions of a model. Multiple replications provide better estimates and more data for confidence intervals.

Results across replications are independent when different random number streams are used, and they are identically distributed because they use the same model logic.

The method of *independent replications* also addresses the problem of autocorrelation by using multiple runs starting from initial conditions (normally $time = 0$). For each run, different random number seeds are used (or possibly different initial conditions). The replications are then independent of each other, and each replication mean is considered an independent observation. Mean, variances, and confidence intervals can be computed from data produced by multiple replications.

The *batch means* method applies to steady-state analysis and avoids the problem of handling transitional periods during independent replications. Batch means use a single run divided into multiple intervals, with the data collected in each interval counted as a batch. For each batch, the mean of a variable of interest is computed. The batch mean values then serve as individual observations for which the grand mean and variance are computed using the same formulas used for independent replications. Data dependency among batch means is significantly weaker than among individual observations in a single batch for a run.

One problem with the batch means method is identification of the length of the interval representing the batch. Shorter intervals result in stronger dependencies between the batch means, because the chance of a given system state during one interval continuing into the next interval is increased. For example, if there are five entities in the system during the ith interval, the same five may remain in the system during the $(i+1)$th interval. This results in a strong dependency among the data (for example, number in queue, waiting time) collected for the two batches. Because of this, larger batch sizes and fewer batches (approximately 5 to 30) are generally recommended.

6.5 OUTPUT ANALYSIS

6.5.1 DESCRIPTIVE STATISTICS

Simulation runs are used to derive estimates of actual population quantities based on the run samples. The primary simulation statistics for this are listed in Table 6.3. This nomenclature is used in subsequent sections for analyzing simulation output.

TABLE 6.3: Simulation Statistics

Statistic	Description
n	Number of simulation observations.
$\bar{X}$	The sample mean of a random variable from simulation based on n observations $x_1, x_2, \ldots x_n$.
S^2	The sample variance of a random variable from simulation based on n observations $x_1, x_2, \ldots x_n$.

(continued)

TABLE 6.3: Simulation Statistics (*continued*)

Statistic	Description
μ	The actual population mean being estimated.
$\sigma_{\bar{x}}$	The actual population standard deviation.
CI	A Confidence Interval (CI) range of values that are likely to contain the true value of the unknown population parameter at a given confidence level.

6.5.2 CONFIDENCE INTERVALS

We desire to know the true value of a simulation output parameter based on the results of many observations. The accuracy of the statistical estimate is expressed as a range called a *confidence interval.* The upper and lower bounds define a range of values where the true value of an estimated parameter is expected to exist. A confidence interval (e.g. for a mean value) is produced from multiple runs and it can be updated after the result of each simulation run. The *confidence level* used in hypothesis testing is the probability that the interval contains the true value of the parameter.

The interpretation of a confidence interval follows: if one constructs a very large number of $(1-\alpha)$ confidence intervals each based on n observations, in the long run $(1-\alpha)$ of the intervals contain the parameter value. The parameter α is also called the significance level, or the probability of rejecting a hypothesis when it is true.

A narrow confidence interval indicating less uncertainty is preferred as a better estimate of a random variable. Conversely, an interval with a high confidence level indicating higher probability is preferred over one with a low confidence level. For given sample data, a confidence interval with a higher confidence level will be wider than a confidence interval with a lower confidence level. Generally, analysts select a confidence level before running simulation experiments.

The choice of sample size affects the confidence interval width and level. A confidence interval derived from a larger sample will be usually be narrower for a given confidence level. Knowledge of the desired interval width and confidence level for the parameter to be estimated can be used to specify the sample size before model execution. One can use the desired confidence interval width in conjunction with estimated sample variance to determine an appropriate sample size.

6.5.3 ESTIMATION OF A POPULATION MEAN

An important measure in a simulation model is the mean value of a variable. Mean estimates are often the focus of analysis such as average waiting time in a queue, average time spent in the system, average length of a queue, or average utilization. Another important simulation output for a variable is its standard deviation indicating the statistical spread (dispersion) around the mean.

If the individual observations of a random variable are denoted by $x_1, x_2, ..., x_n$ for n observations (e.g., waiting times in the system for n entities), the sample mean and variance are calculated using Equations 6.1 and 6.2:

$$\overline{x} = \frac{x_1 + x_2 + ... + x_n}{n} = \frac{1}{n}\sum_{i=1}^{n} x_i \tag{6.1}$$

$$S^2 = \frac{1}{n-1}\sum_{i=1}^{n}(x_i - \overline{x})^2. \tag{6.2}$$

Due to the *central limit theorem*, the distribution of the sample mean can be assumed normal for analyzing simulation output. The central limit theorem states that when a random sample is drawn from any population with mean μ and standard deviation σ, its sample mean $\bar{x}$ has a sampling distribution whose shape is approximately normal as long as the sample size is large enough. The larger the sample used, the more closely the normal approximates the sampling distribution for the mean.

We assume the observations used in the mean and variance Equations 6.1 and 6.2 are independent of one another and identically distributed. Hence the distribution of the mean of the observations, $\overline{x}$, is normal for sufficiently large sample sizes greater than 30 based on the central limit theorem.

To derive a confidence interval based on the normality assumption we define the random variable Z. It is distributed normally with a mean of zero and standard deviation of one as calculated per Equations 6.3 and 6.4.

$$Z = \frac{\bar{x} - \mu}{\sigma_{\bar{X}}} \tag{6.3}$$

$$\sigma_{\bar{x}} = \frac{\sigma}{\sqrt{n}} \tag{6.4}$$

where

$\bar{X}$ is the sample mean
μ is the population mean being estimated
σ is the standard deviation of the population of random variables

A confidence interval for the mean is defined as an interval with a probability $(1-\alpha)$ of containing the true mean. It lies between two symmetric values given by a standard normal distribution table. The interval refers to the central $(1-\alpha)$ portion of the standard normal distribution defined in Equation 6.3 and shown graphically in Figure 6.8. $Z_{\alpha/2}$ is the value of the standard normal random variable that cuts off $\alpha/2$ portions of the symmetric distribution in the tails. This $\alpha/2$ value is referred to as the critical point of the distribution shown in Figure 6.8 (and used in hypothesis testing as described in Section 6.5.5).

The $(1-\alpha)$ confidence interval relationship can be written as

$$P(-Z_{\alpha/2} \leq Z \leq Z_{\alpha/2}) = 1-\alpha. \tag{6.5}$$

Substituting Z in the previous expression gives the confidence interval as follows:

$$P\left(\bar{X} - \frac{Z_{\alpha/2}\sigma}{\sqrt{n}} \leq \mu \leq \bar{X} + \frac{Z_{\alpha/2}\sigma}{\sqrt{n}}\right) = 1-\alpha. \tag{6.6}$$

Figure 6.8 depicts the confidence interval as a normal distribution for the Z test statistic with a mean of zero. The Z-score indicates the number of standard deviations from the zero mean. The true mean falls in the confidence interval with a probability of $(1-\alpha)$. When calculating confidence intervals and in hypothesis testing, use the normal distribution tables for Z if there are greater than 30 samples (which should be strived for). For small sample size less than 30 use the Student t distribution instead to account for the uncertainty associated with the smaller sample.

In hypothesis testing for a mean, the null hypothesis is accepted when the test statistic is in the middle $(1-\alpha)$ portion. The hypothesis is rejected when it falls in one of the critical regions. See Section 6.5.5 for more details of hypothesis testing.

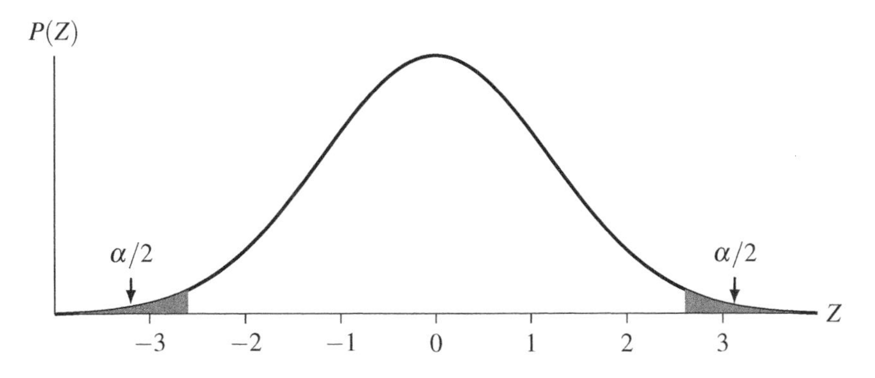

FIGURE 6.8: Confidence Interval

Example: Confidence Interval

The charging station model was run for a duration of 12 hours with uniform distributions to determine a 95% confidence interval for the mean waiting time. Figure 6.9 shows the ending portion of run output with the summary statistics used to compute the 95% confidence interval. Since more than 30 samples are captured, the use of a normal distribution table is justified to find the Z value.

```
Time   Event
  .
  .
  .
711.4   Car 90 Arriving at station
711.4   Car 90 Entering a charging bay
717.4   Car 91 Arriving at station
719.6   Car 90 Charged and leaving
719.6   Car 91 Entering a bay

Cars Entered =  91
Cars Served =  90
Mean Waiting Time = 3.1 minutes
Waiting Time SD = 2.8 minutes
```

FIGURE 6.9: Confidence Interval Run

Using these values in Equation 6.6 for $n = 90$ cars through the system and a Z value of 1.96 per statistical tables for $\alpha = 1 - 0.95 = 0.05$, the confidence interval is computed as:

$$\bar{X} - \frac{Z_{\alpha/2}\sigma}{\sqrt{n}} \leq \mu \leq \bar{X} + \frac{Z_{\alpha/2}\sigma}{\sqrt{n})}$$
$$3.1 - \frac{1.96*2.8}{\sqrt{90}} \leq \mu \leq 3.1 + \frac{1.96*2.8}{\sqrt{90}}$$
$$3.1 - .578 \leq \mu \leq 3.1 + .578$$
$$2.52 \leq \mu \leq 3.68.$$

It is determined that the true population mean lies in the confidence interval range between 2.52 and 3.68 with a probability of 95%.

Other measures with confidence intervals include estimation of the difference between means, estimation of proportions, and others. Estimating the difference between means is shown in the next section.

6.5.4 ESTIMATION OF A DIFFERENCE BETWEEN TWO MEANS

When different scenarios of a system are simulated, the means of measures of effectiveness can be compared. The system alternative with the more desirable measure of effectiveness can be recommended if the difference between the two means is significant. The basis for this statistical method is that the difference between two independent, normally distributed random variables is a normally distributed random variable.

If the means of the samples in each simulated scenario are normally distributed, such as for large enough sample sizes where the central limit theorem holds, then the difference between means will also be normally distributed with the population mean and population variance per:

$$E[\bar{x_1} - \bar{x_2}] = \mu_1 - \mu_2 \tag{6.7}$$

$$Var[\bar{x_1} - \bar{x_2}] = \frac{\mu_1^2}{n_1} - \frac{\mu_2^2}{n_2}. \tag{6.8}$$

Normality of the difference of means implies the next equation:

$$P\left[-Z_{\alpha/2} \leq \frac{(\bar{x_1} - \bar{x_2}) - (\mu_1 - \mu_2)}{\sqrt{\sigma_1^2/n_1 - \sigma_2^2/n_2}} \leq Z_{\alpha/2}\right] = 1 - \alpha. \tag{6.9}$$

The confidence interval results from transforming the previous equation as:

$$P\left[(\bar{x_1} - \bar{x_2}) - Z_{\alpha/2}\sqrt{\frac{\sigma_1^2}{n_1} - \frac{\sigma_2^2}{n_2}} \leq \mu_1 - \mu_2 \leq (\bar{x_1} - \bar{x_2}) + Z_{\alpha/2}\sqrt{\frac{\sigma_1^2}{n_1} - \frac{\sigma_2^2}{n_2}}\right] = 1 - \alpha. \tag{6.10}$$

Example: Comparing Systems by Estimating a Difference between Means

Two of the electric car charging system configurations will be compared by estimating the difference of their means per Equation 6.10. Specifically it will be determined whether adding a third charging bay makes a significant difference in mean waiting time. Figures 6.10 and 6.11 show Monte Carlo simulation summaries of the two configurations using exponential distributions for arrival rate and charging time.

```
Scenario: Capacity = 2, 12 Hours, 1000 Iterations
Grand Mean Waiting Time = 0.914
Mean Waiting Time SD = 0.601
```

FIGURE 6.10: Monte Carlo Results for Two Charging Bays

```
Scenario: Capacity = 3, 12 Hours, 1000 Iterations
Grand Mean Waiting Time = 0.120
Mean Waiting Time SD = 0.130
```

FIGURE 6.11: Monte Carlo Results for Three Charging Bays

Plugging the results into Equation 6.10 gives:

$$(0.914 - 0.120) - 1.96\sqrt{\frac{0.601^2}{1000} - \frac{0.130^2}{1000}} \leq \mu_1 - \mu_2 \leq (0.914 - 0.120) + 1.96\sqrt{\frac{0.601^2}{1000} - \frac{0.130^2}{1000}}$$

$$0.794 - 1.96 * 0.0194 \leq \mu_1 - \mu_2 \leq 0.794 + 1.96 * 0.0194$$

$$0.756 \leq \mu_1 - \mu_2 \leq 0.832.$$

As expected, the confidence interval shows that three charging bays is likely to have a lower average waiting time than two bays. The interval indicates that the waiting time difference is statistically significant. Though the interval is less than one minute difference, it is judged noticeable for an average customer. The charging manufacturer will have to trade off lower waiting time (probably resulting in more customers) vs. the added expense for an extra bay. Repeating this for three vs. four bays produces a very small difference interval of a few seconds, which may be judged as a negligible effect and not worth the extra cost of another charger.

6.5.5 HYPOTHESIS TESTING

Hypothesis testing is used when it is desirable to know if the mean of an output variable is equal to, less than, or greater than a specific value. It can also be used to determine if there is a significant difference between two means obtained from different simulation experiments. Hypothesis testing is based on the assumption of normality of sample means like the previous methods for establishing confidence intervals.

In hypothesis testing one starts with a null hypothesis, H_0, which states a certain relationship that may or may not be true. One also sets the *significance level*, α, which is the probability of rejecting a hypothesis when it is true. The confidence level is $(1 - \alpha)$ as shown previously. The hypothesis is not rejected if the relationship holds true statistically.

Rejecting a null hypothesis that is actually true is called a Type I error. α represents the maximum acceptable risk of making such an error. Accepting H_0 when it is actually false is called a Type II error.

The hypothesis test is performed on the information drawn from a statistical sample of simulation output data. Suppose the null hypothesis H_0 states that the population mean $\mu = a$. If the mean of the random variable x is based on sample of size n, we would fail to reject H_0 under a confidence level of $(1 - \alpha)$ if the following is true:

$$-Z_{\alpha/2} \leq \frac{\bar{x} - a}{S/\sqrt{n}} \leq Z_{\alpha/2}. \tag{6.11}$$

The null hypothesis is rejected if this relationship does not hold.

We fail to reject the null hypothesis stating that $\mu > a$ under a confidence level of $(1-\alpha)$ if this relationship holds true:

$$\frac{\bar{x}-a}{S/\sqrt{n}} < Z_\alpha. \tag{6.12}$$

Similarly, we fail to reject the null hypothesis stating that $\mu < a$ under a confidence level of $(1-\alpha)$ if:

$$\frac{\bar{x}-a}{S/\sqrt{n}} > -Z_\alpha. \tag{6.13}$$

The population standard deviation σ may be replaced by the sample standard deviation s for sample sizes greater than 30. The critical values of Z are again found in standard normal distribution tables.

There is an intimate relationship between the confidence interval in Equation 6.6 and the hypothesis test in Equation 6.11. It can be observed that rejection of the null hypothesis H_0 that $\mu = a$ is equivalent to a not being inside the confidence interval for μ.

For comparing the results of two simulation scenarios, the previous relationships can be modified to test hypotheses about differences between two means. The null hypothesis can be stated that one variable mean is equal to another variable mean, $\mu_1 - \mu_2 = 0$. Equation 6.11 for testing a single variable mean can now be applied to the difference between the two means. For this case we fail to reject the hypothesis if the following is true:

$$-Z_{\alpha/2} \leq \frac{(\bar{x_1}-\bar{x_2})-0}{\sqrt{S_1^2/n1+S_2^2/n2}} \leq Z_{\alpha/2}. \tag{6.14}$$

Example: Hypothesis Testing to Compare Systems

For this example we will continue with the same charging station model scenarios as in Section 6.5.4 for comparison. We want to know whether the difference between the mean waiting times for 2 vs. 3 charging bays under the same conditions is significant. We will set a confidence level of 95%. The null hypothesis H_0 is that there is no significant difference between the two means. The test statistic based on the given data using Equation 6.14 is

$$Z = \frac{(.914-.120)}{\sqrt{0.601^2/1000+0.130^2/1000}} = -4.53.$$

Since the computed Z value is substantially outside the range $[-1.96, 1.96]$ per a normal distribution table for a 95% confidence level, we reject the null hypothesis and conclude that there is a significant difference between mean waiting times due to the system configurations.

6.5.6 SENSITIVITY ANALYSIS

Sensitivity analysis is a useful technique for understanding and using the results of simulation models, and is applicable to all types of models. It explores the effects on key result variables of varying selected parameters over a plausible range of values. This allows the modeler to determine the likely range of results due to uncertainties in key parameters. It also allows the modeler to identify which parameters have the most significant effects on results, suggesting that those be measured and/or controlled more carefully. As a simple example, if a 10% increase in parameter A leads to a 30% change in a key result variable, while a 10% increase in parameter B leads to only a 5% change in that result variable, one should be somewhat more careful in specifying or controlling parameter A.

Designed experiments should be used to analyze model outputs, finding the inputs on which each output is most dependent. Analysis of variance (ANOVA) is the statistical technique ordinarily used to identify the relative influence of each input on each output. A 2^k factorial design is an economical design strategy for larger models and is often used to screen significant factors from insignificant factors. Sensitivity analysis can be performed by varying the parameters manually, or some simulation tools (see Appendix A) automate sensitivity analysis as an advanced feature.

6.5.7 RESPONSE SURFACES

A simulation model is a mechanism that turns inputs into output performance measures; in this sense a simulation is just a function. In many cases a 3-D response surface can represent the output space of a simulation model. Generating visualized response surfaces is thus a useful approach for output analysis. When interesting interactions are found between input variables for a specific output variable, generating data for and plotting a response surface can be instructive. Response surface methodologies are used to seek optimal system configurations. Many of the techniques use a gradient-based approach to find the optimums.

Example: Waiting Time Response Surface

An example response surface will be generated with the electric car charging model. Figure 6.12 shows a surface plot of waiting time by varying mean arrival rate of cars from 20 to 100 per hour and capacity from 2 to 10 charging stations. The interacting effects of the independent variables can be easily visualized including local and global optimas to minimize waiting time. The subset region in the figure covers mostly acceptable operating conditions, whereas the entire decision space would require much greater z-range to display inordinately long waiting times.

6.6 PRESENTATION

As a simulation model is developed and verified, the modeler should be considering ways of presenting results. Success of the entire modeling process requires an under-

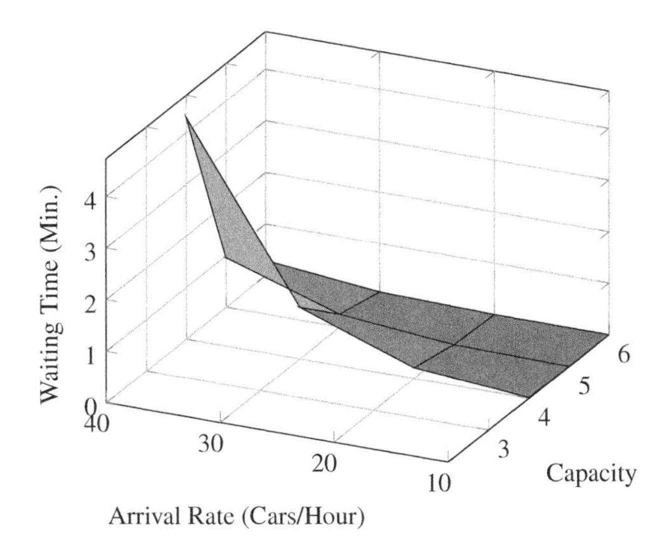

FIGURE 6.12: Waiting Time Response Surface

standable presentation to others and definitive recommendations. All recommendations made should also be based on actual opportunities to manipulate controllable variables.

Models usually require substantial time and effort to construct and verify. Similarly, a model should be exercised extensively to produce results of interest to stakeholders. Answering stakeholders' questions, performing sensitivity analyses with designed experiments, and running various scenarios all provide opportunities for model experiments.

Use graphics to convey significant findings and to produce pictures to which stakeholders can relate. Pictures require explanation but provide much more information than numbers alone. Some plots of interest may include confidence intervals comparing means from different scenarios, two-dimensional plots including trends and probability plots, response surfaces that show interactions between inputs, and contour plots and maps that show relative values.

Use the results of a sensitivity analysis to discuss the most influential input variables in a system or process. The ability to explain the relative influence of inputs gives stakeholders confidence in recommendations from a study.

If results are counterintuitive, trace through the model and understand its dynamics, then explain the model behavior to stakeholders and discuss with them implications for the real system.

6.7 SUMMARY

Virtually all systems and processes of concern are stochastic with randomly varying behavior. Variation in the outputs is the basis for statistical analysis of outputs.

Choosing what kind of runs to make and how many of each requires careful consideration of several related factors. These include the overall consideration of achieving independent samples, run length, transient effects and steady-state periods, the number of runs, and settings of input variables for controlled experimentation. Whether systems are terminating or non-terminating, or in transient or steady state, has impact on the analysis methods used.

In experimental design we decide what configurations to run before simulation. These experiments should be carefully thought out in order to minimize the number of runs and get the maximum information from the simulations. Factors are chosen to be run at different values. A 2^k factorial design (see Section 6.4.1) is again an economical design strategy, especially for larger models, and is often used to screen significant factors from insignificant factors.

Simulation is a sampling method in itself and correspondingly its output is subjected to statistical analysis. We would like to determine the characteristics of specific outputs with known confidence. Statistical methods are used to determine the confidence intervals for output parameters and support hypothesis testing on the values of the estimated parameters.

Important measures in simulations are the mean value of a variable and its sample variance (dispersion). We desire to know the true value of a simulation output parameter based on the results of many observations. The accuracy of the statistical estimate is expressed as a range of values called a confidence interval that is likely to contain the true value of the unknown population parameter at a given confidence level.

Hypothesis testing is used to assess the mean values of output variables. It can also be used to determine if there is a significant difference between two means obtained from different simulation experiments, such as when comparing system configurations.

Sensitivity analysis explores the effects on key result variables of varying selected input parameters. It helps to identify which parameters have the most significant effects on results. A 3-D response surface visualization is handy to represent the output space of a simulation model.

Presentation of results is important to convey output results to all stakeholders and instill confidence. Graphics and plots of the intervals can impart more information than tables. Some of the techniques for sensitivity analysis and response surfaces provide easily understood visualization.

7 Engineering Case Study Example

7.1 INTRODUCTION AND OVERVIEW

This chapter illustrates the modeling process with a case study conducted in the field of space systems product development. Systems developed for space applications, whether for ground systems or for satellites, are often large, complex, and software-intensive. These systems are developed for missions critical to both government and commercial interests and have to meet schedules for deployment. Static tools, such as size estimation algorithms, Program Evaluation Review Technique (PERT) charts, and Gantt charts are typically used to produce schedules, but they lack feedback loops for representing the dynamic behavior of rework cycles. Thus simulation, with its ability to represent dynamic behavior in product development, was indicated.

7.2 BACKGROUND

This study started when a program office member asked when the satellite software would be ready. The code had been written and integrated and was entering a system testing phase. The program office doubted that the testing could be performed within the time allotted by the contractor, which was less than a year, but could offer no better estimate. So the program office turned to a modeler and asked whether he knew of a way to ascertain when the software would be ready for flight. The question and answer were critical because a launch had to be scheduled. So the question that required an answer was "When will the software be ready?"

The modeler's domain knowledge helped because he had learned, from study of product development modeling, that rework is inherent in product development and that the critical behavior that had to be represented was the rework cycle. Modelers in the field of system dynamics [4] have demonstrated the centrality of the rework cycle in representing project behavior in product development. The modeler also, through his own software development experience, knew that software development is described well by the rework cycle, and that even software developers speak of test-and-fix cycles.

The software test-and-fix cycle (TaFC) is a case of the rework cycle in which software defects introduced during specification, design, and implementation have not been discovered as the product enters testing. Despite advances in software development processes, the TaFC remains a development phase of central importance for complex software-reliant systems. In extreme cases, lack of quality-inducing effort prior to software or system integration testing produces poor-quality software and leads to a lengthy test-and-fix (TaF) phase.

A simple analysis of a TaFC in Figure 7.1 is provided here as a basis for understanding the role of modeling and simulation in addressing this problem. Assuming all test cases are available (the process is not supply constrained), the duration of this process, calculated per Equation 7.1, is measured from the time the first test case enters testing to the time the last test case exits the process. It is dependent on the capacity of the process as in Equation 7.2.

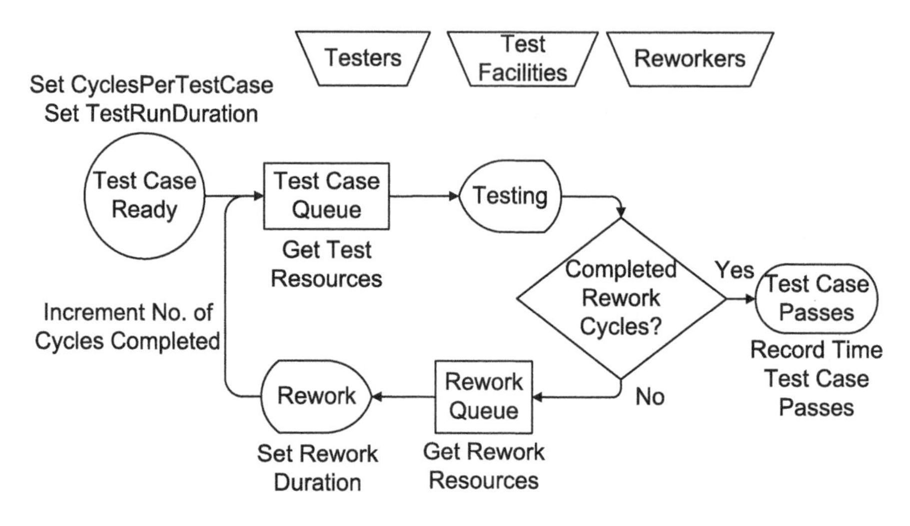

FIGURE 7.1: A Simple Test-and-Fix Model

$$TaF\,duration \approx \frac{\#\,of\,test\,cases * \#\,of\,rework\,cycles/test\,case}{process\,capacity} \tag{7.1}$$

Process capacity is constrained by the resources available in each activity. Specifically, it is determined by the activity having the least capacity per Equation 7.2.

$$TaF\,process\,capacity = MIN(testing\,capacity, rework\,capacity) \tag{7.2}$$

Equation 7.3 illustrates the simplest case using constant values: 200 test cases each requiring 3 rework cycles, with the constant resources and activity durations in Table 7.1. The resulting 1200 hours are equivalent to 30 weeks at 40 hours/week.

TABLE 7.1
Example Values for TaF Duration Calculation

Activity	Resource Units	# Resource Units	Activity Duration	Activity Capacity
Testing	1 tester & 1 testbed	10	2 hrs	5 tests/hr
Fixing	1 software developer	10	20 hrs	.5 reworked case/hr

$$TaF\,duration \approx \frac{200\,test\ cases * 3\,rework\,cycles/test\,case}{MIN(5/hr, 0.5/hr)} = 1200\,hrs \quad (7.3)$$

In actual cases, these simplifying assumptions of constant values do not hold and wide variation in number of rework cycles per test case inhibits an analytical calculation. Estimation of the duration and required resources for rework cycles is inhibited by task size variation, process capacity variation, and the process characteristics of dynamism and concurrency.

This difficulty is especially evident in testing of software-intensive systems, as illustrated by the following scenario. A customer is apprised that system components are being integrated and they are provided a schedule for completion, based on test durations or on best guesses. But integration and testing produces a seemingly never-ending stream of defect reports followed by rework, resulting in a test-and-fix (TaF) phase that endures well beyond the original schedule. The development organization measures progress of the test phase in terms of test cases executed, defects found and fixed, defect backlog reduction, and test cases completed with fixes. However, none of these four measures provides a means of forecasting TaF completion until late in the phase, well after deployment planning is underway. As schedule pressure increases in a typical scenario, staffing is increased, test cases and facilities usage are reprioritized, schedules are overrun and re-planned, and processes are revised. The customer, having become highly skeptical of contractor progress reports and schedules, calls for independent estimates of the project's progress. One of these estimates is produced from a simulation of the TaF phase and produces an accurate timeline that the customer then uses for planning system deployment.

7.3 MODELING

The first questions that a modeler faces when representing system or process behavior concern the elements that must be represented and the changes in their states that must be represented. The type of modeling determines how the elements are represented. For continuous models, flows are represented, so one must consider what flows and how the flows are modified. For discrete event models, one must consider what entities move through what activities and the resources required in those activities.

In the first attempt at modeling TaFCs, the modeler employed an implementation of a published system dynamics model of product development [4] seen in Figure 7.2. Test cases were chosen as the element flowing through the model. Test cases were defined as a combination of tests run on the software, the software function(s) being tested, and the staff required to perform the testing. This definition allowed us to avoid measuring the staff and productivity, concerning ourselves with only the time required for each operation in the process.

The model was configured such that *producing work products* represented the first test using each test case with the software and *reworking work products* represented all subsequent test runs. Discussions with the project managers elicited five parameters of primary concern in the TaF stage: the incoming rate of test cases from the first stage, the availability of test facilities (*Test Facility Availability*), the time required to exercise each test case (*Test Run Duration*), the number of test runs required for each test case to ensure it runs without problems (*No. Test Runs per Test Case*), and the time to fix each problem found in testing (*Fix Time*). Often, test cases must be run many times. Testing is performed for a case until it cannot proceed further, at which time all the discovered defects have to be fixed before the test case can be run again.

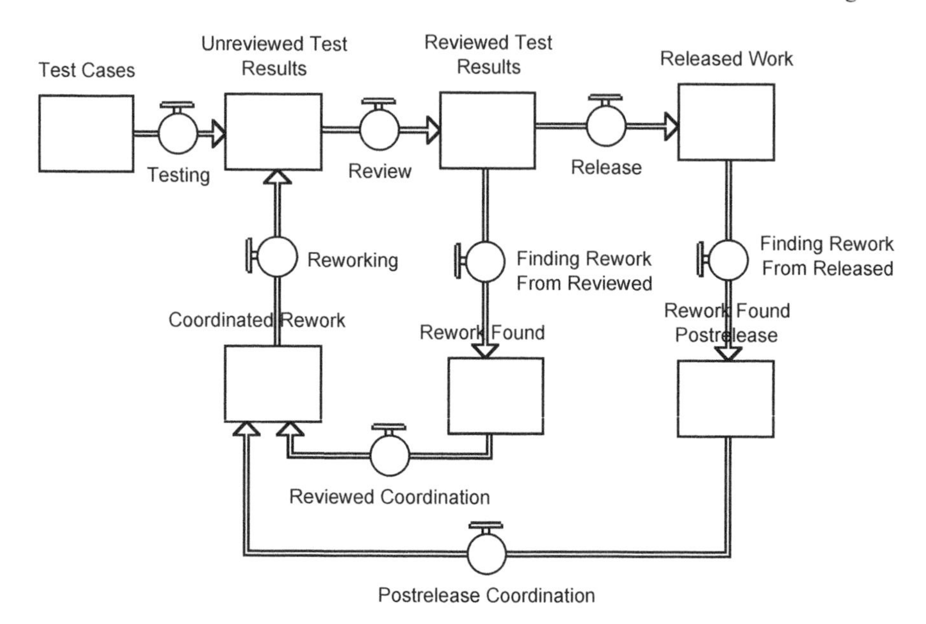

FIGURE 7.2: System Dynamics Model of a Product Development Phase

In the course of configuring the model to the test process, the modeler realized that the model abstracted the process too much to adequately represent the TaF stage of cycling each test case through testing and fixing. Specifically, this model calculates a fraction of work product to be reworked rather than circulating all work products for rework a number of times dependent on the number of defects. The wait times for test cases also needed to be specified and experiments run by varying the input parameters. These requirements are more easily addressed by modeling entities rather than continuous flow. Consequently, a discrete event model for the TaFC was produced.

The discrete event model in Figure 7.1 depicts arriving test cases receiving a value for the attribute, *Runs per Test Case*. Test cases are queued for a test facility and in the queue they are sorted by the number of runs that each has completed so far, giving preference of test facility usage to cases further along in the testing process. The test case at the front of the queue enters the first available facility, staying for the time required to perform the test case. After the test, the number of test sessions for the test case is incremented and checked: if it has reached its required number of runs, it is released for final testing, otherwise it is delayed while a problem is diagnosed and fixed. After the *Fix Time*, the test case is returned to the queue for another test run. Note that in this model, the entity represents a test case during testing but, strictly speaking, it represents the rework discovered by use of the test case during the rework activity. Nonetheless, the representation works due to the close association of a test case and the rework found through its use.

In this model, *Runs per Test Case*, *Test Duration*, and *Rework Duration* are the primary random variables. In each case to which this model has been applied, data was collected from test team leaders by eliciting values for these three variables. The test cases are grouped and for each group, the test leaders are asked to offer minimum, maximum, and most likely values for *Runs per Test Case* and *Test Duration*. One set of *Rework Duration* values is elicited for all test cases because rework duration is not a function of test case type. The elicited values are used to produce random distributions for sampling during model runs. During each run of the model, each test case is assigned a number of runs per test case from the random distribution for its test case group. Each time a test case is represented as being exercised, the test duration is randomly generated from the random distribution of test durations for its test case group. Each time a test case goes for rework, the rework duration is randomly generated from the distribution of rework durations.

The primary output of the model is test case completions over time. The model is run many times and the completion time of each test case is recorded. The completion times are arrayed in a spreadsheet, each column containing results for a run and each row containing results for a test case in the order they are completed. For each row, statistics are calculated: the average completion time with a 95% prediction interval. These are plotted and typically produce an S-curve (Figure 7.3) in which three stages can be distinguished. In the first stage, the number of test case completions accelerates (increasing slope). The process initially fills up with work and the first cases finish slowly but the pace of completions increases. In the second stage, the process is filled with test cases, the process is working at capacity, and the slope of the output line becomes constant. It remains so until the third stage when the process output rate starts declining because less test cases are being processed. The dashed lines in Figure 7.3 represent the 95% prediction interval.

The plot is used to forecast test phase completion, but it is also used to track actual completions. Actual completions are plotted and attention is paid to their relationship to the prediction interval. In practice, actual completions do not produce smooth curves. Completion points may fall outside the interval, either above it or below it, during the test phase and then fall within the interval as the test phase nears completion.

When the first TaFC model was used to forecast test phase completion, the results showed that the test phase would take 2.5 times what the contractor was planning. Knowing this, the customer was able to plan realistically for system deployment.

If a model provides a good representation, providing useful information, it will likely be reused and enhanced. Variations of this model were used for six programs

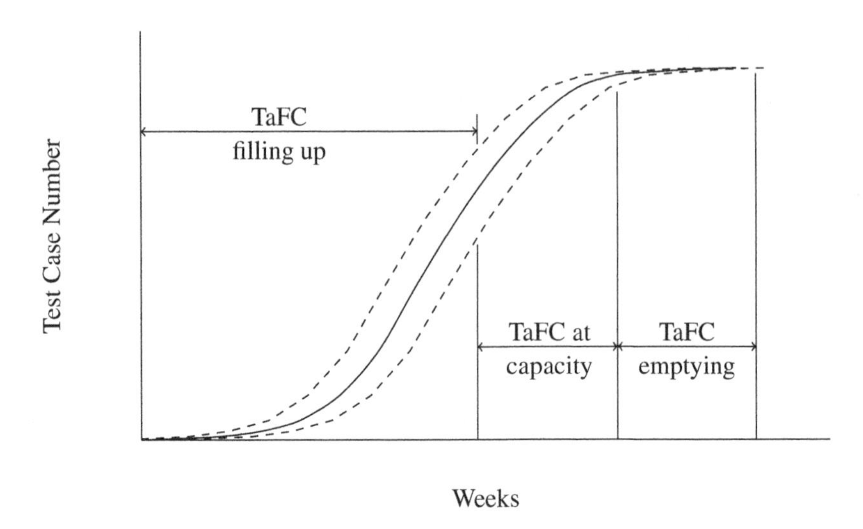

FIGURE 7.3: Example Test Case Completion Progress

before an analysis of commonality and diversity of the models was performed and a generalized test-and-fix model was developed [10]. Two enhancements to this model have been distinguishing types of rework (software fixes versus test case fixes versus hardware fixes) and using two cycles to represent two sequential test phases.

Dynamic modeling of test-and-fix cycles has provided good forecasting of test phase durations, but has also produced good results in supporting resource allocation and process improvement decisions. These results are produced through two types of exercises, sensitivity analysis and scenario analysis, which are discussed next.

7.4 DISCOVERING SIGNIFICANT FACTORS AND BEHAVIOR

Sensitivity analysis can be very beneficial for providing an understanding of how a whole system or process works. This section discusses experiments designed and run to understand the significant factors and behavior in a test-and-fix model. The first experiment performed on the model was designed to determine the relative importance of the modeled input factors. The arrival rate of test cases is often a constant value, in this example, one new test case per day available for testing. For the remainder of the factors, a full factorial (2^4) experiment was run per Table 7.2.

In one case, program office personnel expected *Test Facility Availability* to be the dominant factor in the test-and-fix cycle. However, experiments showed that *Runs per Test Case* was the dominant factor in software test duration, in this example solely contributing to 54% of TaF duration variation and contributing to another 22% through interactions with other factors. In particular, this factor interacts strongly with *Test Run Duration* shown in Figure 7.4.

In a follow-up experiment, additional points were added to the experimental de-

TABLE 7.2
Factors and Values for Screening Experiment

Factor	Low Value	High Value
Test Facility Availability	60 hrs/ week	100 hrs/week
Runs per Test Case	2	8
Test Run Duration	2 hrs	5 hrs
Fix Time	24 hrs	96 hrs

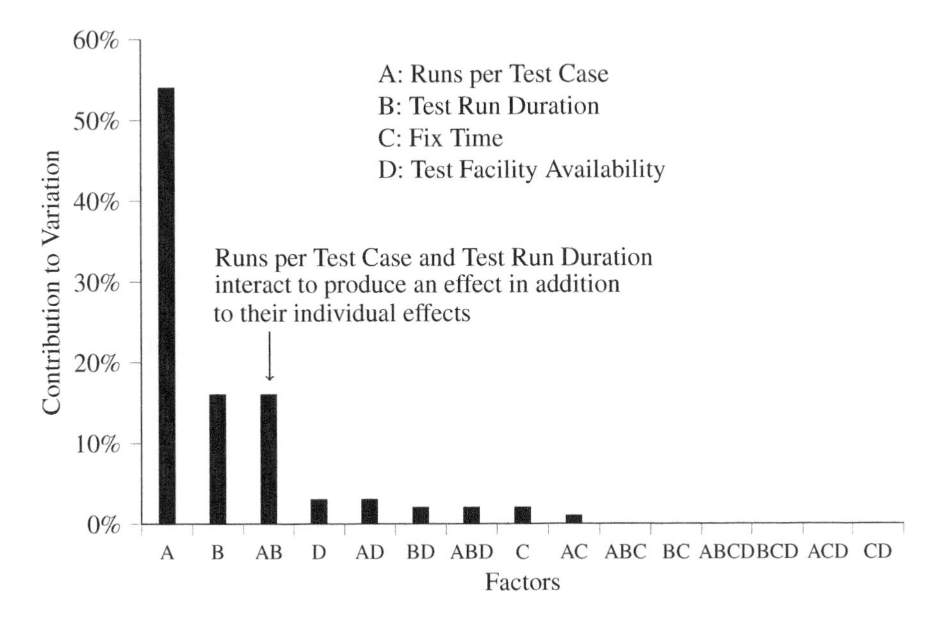

FIGURE 7.4: Contribution to Variation in Software Test-and-Fix Cycle Duration

sign per Table 7.3 in order to obtain response surfaces and a better understanding of the constrained test-and-fix cycle behavior. This experiment considered *Test Facility Availability* at two levels, 60 and 100 hours/week.

Response surfaces produced by this experiment show two regions distinguished by a threshold in Figures 7.5 and 7.6. On one side of the threshold is a planar surface; on the other, a steeply ascending surface rising to the point of the highest values of *Test Run Duration* and *Runs per Test Case.*

The influence of *Runs per Test Case* is evident in both regions. The planar region reflects almost no increased duration due to *Test Run Duration*, but a slowly rising duration due to *Runs per Test Case.* In the steeply ascending region, *Runs per Test Case* interacts strongly with *Test Run Duration.*

TABLE 7.3
Factors and Values for a Response Surface

Factor	Values
Test Facility Availability	60 and 100 hrs/week
Runs per Test Case	2, 4, 6, 8
Test Run Duration	2, 3, 4, 5 hrs
Fix Time	7 days

The threshold represents the points at which the test facilities are fully utilized. When all other factors are constant, the full utilization threshold is a function of test facility availability. Thus, as *Test Facility Availability* decreases, shown in the difference between the two response surface graphs in Figures 7.5 and 7.6, the threshold moves out across the planar region, adding more area to the steeply ascending region and causing it to rise to a higher point. In practical terms, this means that the tighter constraint on test facility availability produces full utilization sooner and the accompanying interaction between *Test Run Duration* and *Runs per Test Case* lengthens the TaF duration.

It became clear from Figures 7.4, 7.5, and 7.6 that test facility availability, though not the dominant factor in duration of the constrained test-and-fix cycle, is active as a facilitator of the dominance and interaction of other factors.

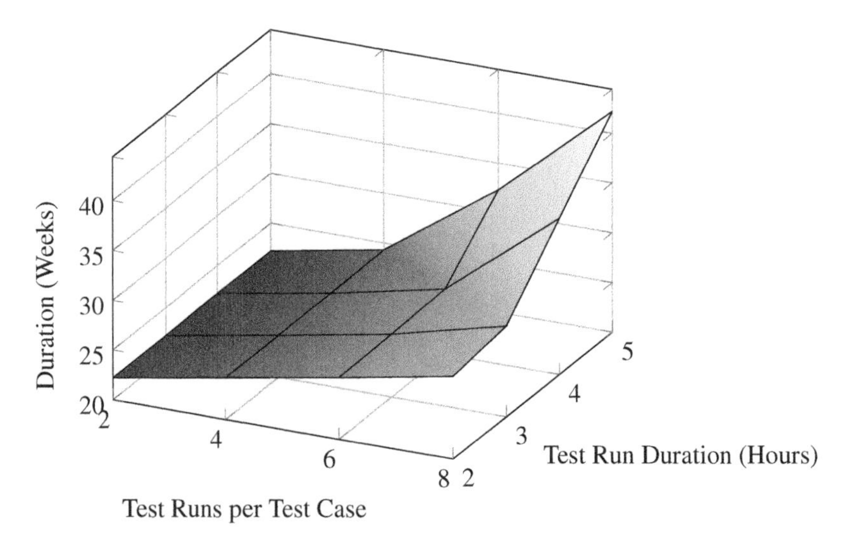

FIGURE 7.5: Response Surface Showing Interaction for Test Facilities Availability of 100 Hours/Week

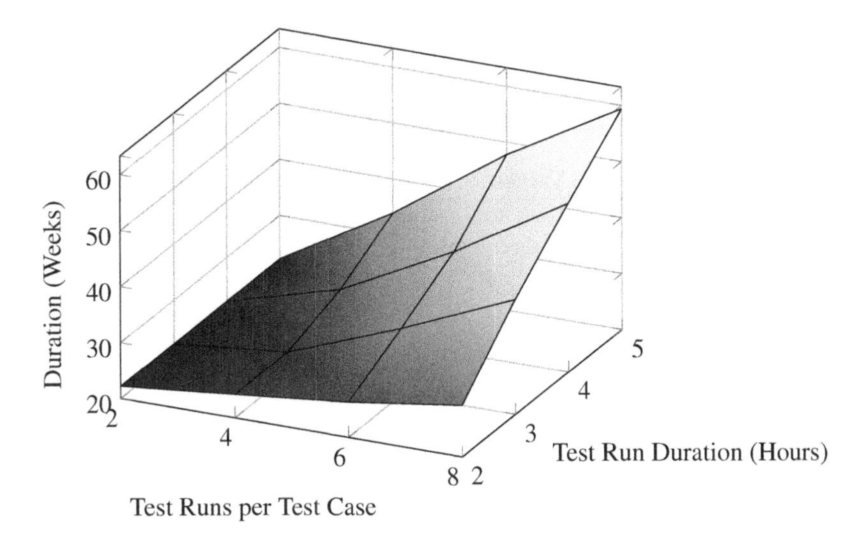

FIGURE 7.6: Response Surface Showing Interaction for Test Facilities Availability of 60 Hours/Week

7.5 MODELING LIKELY SCENARIO AND ALTERNATIVES

Scenario analysis is often performed in response to inquiries made by sponsors of a modeling project. After seeing initial model results, modeling sponsors typically ask what happens when some parameter values are varied in particular ways. For example, what happens if an additional test facility is added or additional testers are made available? In these cases parameter values are varied, results are generated and compared, any unexpected results are analyzed to understand the behavior, and once understood, the results are reported to the inquirers.

The next experiment used likely inputs to estimate the duration of the test-and-fix cycle per Table 7.4. In this example, the project planned to use each test facility 40 hours per week. The number of runs per test case could vary widely depending on staff experience, process maturity, etc., but experience suggested a discrete distribution in which some cases required only 2 runs, a few might require as many as 8, and the most likely number of runs per test case was 4. The time required to run individual test cases, including the amount of test facility time required for setup and documentation, was modeled as a continuous triangular distribution between 2 and 5 hours with a mode of 3.5. Experience in previous projects showed that fixes normally required between one and two weeks, so *Fix Time* was modeled as a uniform discrete distribution for this period. Table 7.3 lists the distributions and how they were sampled.

Figure 7.7 shows the results of 10 simulation runs, enough to exhibit both the variation and the pattern of test case completions. Despite the input variations, the completion time curves for 150 test cases produced a clear pattern comprised of two sections: the curved tails and the straight middle. The length of the tails was found to

TABLE 7.4
Inputs for Estimating Test-and-Fix Cycle Duration

Factor	Values	Sample
Test Facility Availability	Both test facilities at 40 hrs/week each	Constant for all simulation runs
Runs per Test Case	(2, .1), (3, .1), (4, .3), (5, .2), (6, .1), (7, .05), (8,.05)	Randomly for each test case in each simulation run
Test Run Duration	Triangular (2, 3.5, 5) hrs	Randomly for each test case in each simulation run
Fix Time	(7, .125), (8, .125), (9, .125), (10, .125), (11, .125), (12, .125),(13, .125), (14, .125) days	Randomly for each test cycle of each test case in each simulation run

be dependent on the *Fix Time* values: as the *Fix Time* distribution increases, it takes longer to initially complete test cases, as well as to clear the last test cases out of the cycle. The straight middle region occurs with full utilization of the test facilities. The slope of this linear region is largely determined by test facility availability, with a limit due to the incoming rate of test cases.

Confidence and prediction intervals are calculated for estimation purposes. For this example, the 95% confidence interval on the mean duration is 32 to 34 weeks (error estimate = .43), and the 95% prediction interval on duration is 29 to 36 weeks.

Project personnel often want to look at ways to reduce the test phase duration, so a number of questions about alternatives were posed and studied using the model. For example, two questions were asked about increasing test facility availability, either through taking time late in the test phase from another test facility, or by constructing a new test facility. Neither of these options significantly reduced TaF duration because the additional testing capacity came too late in the TaF stage.

Sometimes project personnel question the initial assumptions behind their inputs and ask for simulation results using more optimistic assumptions. To illustrate, more optimistic distributions for *Runs per Test Case* and for *Fix Time* were tried in this example and they produced better results: improved *Fix Time* alone resulted in a 12% duration reduction, improved *Runs per Test Case* alone produced a 26% improvement, and improved *Fix Time* and *Runs per Test Case* together produced a 43% improvement. Achieving these results in practice is considered unlikely but they can

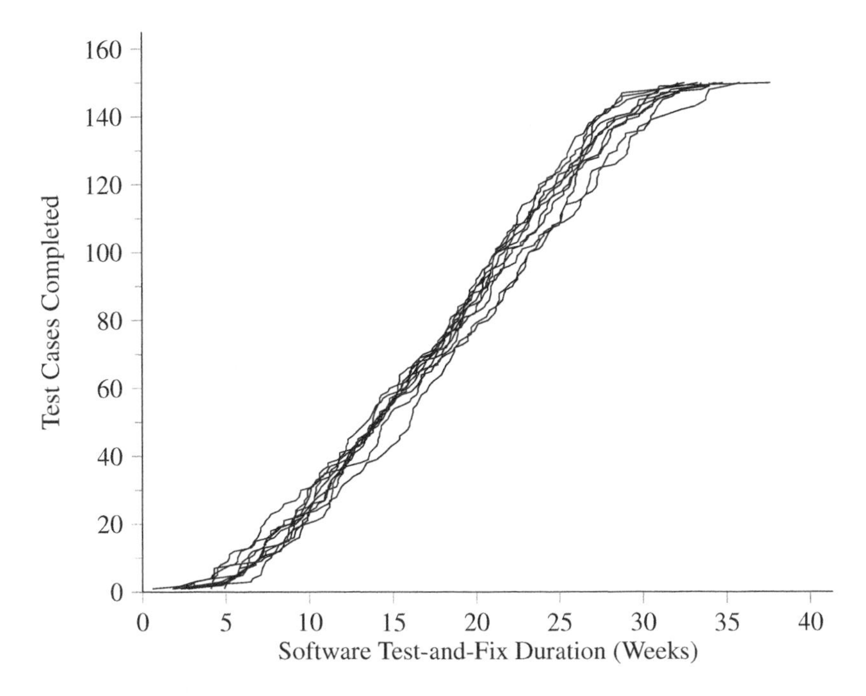

FIGURE 7.7: Test Case Completion Times for 10 Model Runs

demonstrate what is required to achieve a proposed test schedule.

7.6 FINDINGS AND IMPACTS

As mentioned in the first section of this chapter, estimation of a TaF phase duration is very difficult due to the process dynamics. Attempts to assign an average time per test case ignore the important process variables discussed in the previous sections. Also, attempts to solicit expert opinions and aggregate them do not work because the process nonlinearities are usually too challenging for mental calculation. Finally, the parametric software estimation models are usually too high level to produce reasonable estimates of a phase involving significant amounts of rework.

Given this background, the production of a realistic forecast of TaF duration is regarded as an important finding. This information not only allows realistic planning, but it also reduces wasted effort by saving time that would otherwise be spent in discussing whether or not software will be ready when required, and redeveloping plans each time a schedule slips because the previous estimate was unrealistic.

In addition to a good forecast of duration, the modeling process produced process insights that are useful for TaF management. In the preceding example, the modeling allowed identification of the number of test runs per test case, rather than test facility availability, as the dominant factor in test phase duration. Though test facilities certainly enabled earlier completion of the test phase, the more important factor is reducing the number of test runs per test case because limited test resource availability greatly magnifies the effect of this factor once the test facilities become fully utilized. This factor is almost entirely dependent on the quality of the inputs to the testing phase, most importantly the software under test but also the test cases and the test facility.

Furthermore, it was found that the effects of delays and factor interactions are generally not understood and therefore not taken into account when estimating test duration.

Understanding the TaF cycle dynamics enabled recommending a set of guidelines for reducing the duration of this cycle, ordering the first four based on the relative influence of factors. These are elaborated below.

Conduct well-performed quality-inducing activities, for example, peer reviews, unit testing, and defect causal analysis prior to employing the test facilities. The modeling provided case-based, quantitative support for this quality cost principle.

Reduce the number of test sessions. If a test case fails, continue running it as far as possible in order to find further anomalies. Reduce diagnostic time through good anomaly documentation. This trade-off in favor of reducing the number of test runs comes at the expense of the less important factor, test run duration.

Reduce the fixed time of test runs (setup, recordkeeping, saving files). Automation can significantly reduce time in a test facility, as well as facilitating good documentation of anomalies.

Reduce the learning time of testers through training and regular communication. The efficiency of a test team can be undermined by individuals optimizing their personal work at the expense of the team's work, such as failure to plan for shared use of common resources.

Complete TaF before FT. This suggests that showing progress early may carry a cost in the form of a longer overall duration.

As the end of TaF approaches, focus on reducing the time to provide fixes. As fewer test cases are left in the test-and-fix cycle, fix delays can dominate the cycle duration.

Though these guidelines may seem obvious to those well-versed in the economics of software quality, many development groups have only a vague notion of quality costs, and are inclined to choose poorly when working under cost and schedule pressures. Using a simulation model to demonstrate the dynamics of a specific process can make a very persuasive argument for good choices, as well as quantifying the relative importance of factors.

7.7 SUMMARY

This chapter illustrated the modeling process with a case study conducted in the field of space systems product development. It was undertaken to answer a stakeholder question about when the software would be ready.

The case study demonstrates the modeling process steps with examples of important concepts discussed in earlier chapters including model engagement, specification, data collection, modeling, assessment, and making recommendations. The model construction was shown to be iterative with an initial system dynamics model of product development, and eventual construction of a discrete event model to better address the model purpose.

Dynamic modeling of test-and-fix cycles provided good forecasting of test phase durations, and made important impact by supporting resource allocation and process improvement decisions. With the results of the modeling, the customer was able to plan realistically for system deployment.

The results included sensitivity analyses and scenario analyses. The experience showed that sensitivity analysis can be very beneficial for providing an understanding of how a whole system or process works. Scenario analysis was necessary in response to inquiries made by the modeling project sponsors.

The modeling project was able to answer the original question and gave additional organizational benefit. The modeling produced a realistic forecast of test-and-fix duration, and provided process insights useful for test management. Overall, this case study experience exemplified the general methods, processes, and benefits of M&S.

Appendix A Simulation Tools

This appendix lists both commercial and free open-source modeling tools. Most of the tool vendors also have free demonstration versions that are normally limited in their modeling capabilities (e.g. allowing a limited number of model entities). The broad range of modeling approaches, languages, and platforms offer many choices of simulation tools. The model construction capabilities may be visual, textual programmed languages, or a combination.

The following tools listed have a wide variety of features for engineering applications. The ones below are recommended as well-known tool vendors with long-established user bases and more recent open-source solutions. They are a combination of desktop and web-based tools that will continue to evolve rapidly.

A.1 COMMERCIAL TOOLS

AnyLogic

AnyLogic is a multi-method simulation modeling tool for business and science developed by The AnyLogic Company. It supports agent-based, discrete event, and system dynamics simulation methodologies. The AnyLogic PLE edition is available free for educational purposes.
`http://www.anylogic.com/`

Arena

Arena is a discrete event simulation tool. It provides for graphical construction using flowchart modeling methodology with a large library of pre-defined building blocks. Many statistical distributions are available, as well as statistical analysis and report generation, performance metrics and dashboards, 2D and 3D animation capabilities. There are custom versions for industry and academic users. A free trial version is available.
`http://www.arenasimulation.com/`

ExtendSim

ExtendSim is a simulation tool for discrete event, continuous, discrete rate, and agent-based simulation developed by Imagine That Inc. Modeling constructs include graphical building blocks, animation, evolutionary optimization, Monte Carlo, batch-mode, sensitivity analysis, and more. A free limited version is available.
`http://www.extendsim.com/`

iThink

iThink is a system dynamics modeling tool developed by isee systems. Stella is a version for academic purposes They provide graphical construction of models with features for Monte Carlo, sensitivity analysis, animation, customizable interfaces, and more. Limited free versions are available.
`http://www.iseesystems.com/`

Vensim

Vensim is a simulation tool developed by Ventana Systems that primarily supports continuous simulation (system dynamics), with some discrete event and agent-based modeling capabilities. It is available commercially and as a free Personal Learning Edition.
`http://vensim.com/`

A.2 COMMERCIAL DISTRIBUTION FITTING TOOLS

ExpertFit

ExpertFit is distribution-fitting software that will automatically determine which probability distribution represents a data set. It has numerous distributions, graphical plots, goodness-of-fit tests, sample sizes up to 100,000, interactive histograms, support for simulation modeling, a distribution viewer, and batch mode.
`http://www.averill-law.com/distribution-fitting/`

StatAssist

StatAssist is a tool to explore the properties of probability distributions. It works with the EasyFit data analysis and simulation application to fit distributions to sample data and apply the results. It supports numerous continuous and discrete distributions, has automated fitting combined with manual fitting, interactive graphs, goodness-of-fit tests, and random number generation. There is a free trial version.
`http://www.mathwave.com/help/easyfit/index.html`

A.3 OPEN SOURCE AND NO-COST TOOLS

Insight Maker

Insight Maker is a free, open-source modeling and simulation tool developed using web-based technologies and supports graphical model construction using multiple paradigms [7]. System dynamics models and agent-based models can be developed, executed, and shared online.
`http://InsightMaker.com`

JaamSim

JaamSim is a free and open source discrete event simulation software that includes a drag-and-drop user interface, interactive 3D graphics, input and output processing, and model development tools and editors [15]. It is based on the Java language.
`http://jaamsim.com/`

NetLogo

NetLogo is a multi-agent programmable modeling environment. It comes with a large library of sample models, and includes a series of multi-agent modeling languages including StarLogo and StarLogoT. It is under continuous development and free of charge.
`http://ccl.northwestern.edu/netlogo/`.

Repast

The Recursive Porous Agent Simulation Toolkit (Repast) Suite is a family of free, and open source agent-based modeling and simulation platforms under continuous development [1]. It has multiple implementations in several languages [22] and built-in adaptive features, such as genetic algorithms and regression.
`http://repast.sourceforge.net/`

SimPy

SimPy is a discrete event simulation library written in the Python language [27]. It is open source with extensive examples and an active user community.
`http://pypi.python.org/pypi/simpy`

Appendix B Simulation Source Code Examples

The source code files in this appendix and more are maintained in a GitHub repository available at https://github.com/madachy/What-Every-Engineer-Should-Know-About-Modeling-and-Simulation. There, the reader can find updates, variations of the programs, and other model examples to download or clone. Links to web-based simulation models are also provided.

B.1 DISCRETE EVENT SYSTEM MODELS

The programs below are variations of the electric car charging station model demonstrated in the chapters. They use the open-source SimPy library for discrete event simulation models [27] with the Python language. At the time of this writing SimPy 3.0 was available but without the full monitoring features of SimPy 2.0. Both versions were used in this book and are identified as such.

B.1.1 SIMPLE MODEL FOR ELECTRIC CAR CHARGING STATION

The model listed below simulates the electric car charging queue/server with fixed times from Chapter 1 using the SimPy 3.0 library. It can be executed in a web browser at https://repl.it/HJBY/3.

```
# simulation of electric car charging station

import simpy
import statistics

# instantiate execution environment for simulation
environment = simpy.Environment()

# charging station resource
charging_station = simpy.Resource(environment, capacity
    =1)

# fixed interarrival and charging times
interarrival_times = [2, 8, 7, 2, 11, 3, 15, 9]
charging_times = [11, 8, 5, 8, 8, 5, 10, 12]

# statistics initialization
waiting_times = []
```

```
# electric car process generator
def electric_car(environment, name, charging_station,
    arrival_time, charging_time):
      # trigger arrival event at charging station
      yield environment.timeout(arrival_time)

      # request charging bay resource
      print('%s arriving at %s' % (name, environment.now
          ))
      with charging_station.request() as request:
          yield request

          waiting_time = environment.now - arrival_time
          print('%s waiting time %s' % (name,
              waiting_time))

          # charge car battery
          print('%s starting to charge at %s' % (name,
              environment.now))
          yield environment.timeout(charging_time)
          print('%s leaving at %s' % (name, environment.
              now))

          # collect waiting times
          waiting_times.append(waiting_time)

arrival_time = 0
# simulate car processes
for i in range(7):
      arrival_time += interarrival_times[i]
      charging_time = charging_times[i]
      environment.process(electric_car(environment, 'Car
          %d' % i, charging_station, arrival_time,
         charging_time))
      #print('%s waiting time = %s' % (i, waiting_time))

environment.run()
print ("Mean Waiting Time = %3.1f minutes" % statistics
    .mean(waiting_times))
print ("Variance of Waiting Time = %3.1f minutes" %
    statistics.variance(waiting_times))
```

B.1.2 FULL MODEL FOR ELECTRIC CAR CHARGING STATION

The model below enhances the simple model for randomness with multiple probability distributions, Monte Carlo simulation, and additional reporting features. It was used for most of the demonstration examples in Chapters 3 through Chapter 6. It is fully generalized to replicate all the examples in this book, uses the SimPy 2.0 library with the full monitoring features, and is an example of object-oriented programming. With switches in the code one can use different distributions, choose Monte Carlo options, and enable other options. It can be easily extended. A version of it can be executed at `https://repl.it/FPxV/72`.

```
# simulation of electric car charging station

from SimPy.Simulation import *
from random import Random,expovariate

import statistics
import math

# electric car process
class Electric_Car(Process):
  def __init__(self,name,charging_time):
     Process.__init__(self,name=name)
     self.charging_time = charging_time

  def visit_station(self):
     global cars_served
     if monte_carlo == False: print ("%4.1f  %3s
        Arriving at station" % (now(), self.name))
     arrive = now()
     yield request,self,charging_station
     if monte_carlo == False: print ("%4.1f  %3s
        Entering a charging bay" % (now(), self.name))
     waiting_time = now()-arrive
     car_monitor.observe(waiting_time)
     yield hold,self,self.charging_time
     yield release,self,charging_station
     cars_served += 1
     if monte_carlo == False: print ("%4.1f  %3s
        Charged and leaving" % (now(), self.name))

simulation_time = 720
#monte_carlo = True
monte_carlo = False
num_iterations = 1000
```

```
if monte_carlo == False: num_iterations = 1

# exponential or uniform
distribution_type = 'exponential'
charging_bays=1

global cars_served

average_waiting_times = []

output_file_name = '../'+distribution_type+str(
    num_iterations)+'_'+str(simulation_time)+'
    min_capacity'+str(charging_bays)+'.txt'

output_file = open(output_file_name, 'w')
output_file.write("iteration average_wait utilization"
    + '\n')

# fixed times for testing
# interarrival_times = [2, 8, 7, 2, 11, 3, 15, 9]
# charging_times = [11, 8, 5, 8, 8, 5, 10, 12]

#exponential distribution
mean_interarrival_time = 6.0
mean_charging_time = 5.0

for iteration in range(num_iterations):
  initialize()
  car_monitor = Monitor()

  charging_station = Resource(capacity=charging_bays,
                         name='charging_station',unitName=
                             'charging bay',monitored=True,
                              monitorType=Monitor)
  arrival_time = 0
  cars_served = 0

  if monte_carlo == False: print ('Time  Event')

  # continue loop until events meet simulation time
  for i in range(200):
       if distribution_type == 'exponential':
           arrival_time += random.expovariate(0.16)
           charging_time = random.expovariate(0.2)
```

```
        if distribution_type == 'uniform':
            arrival_time += random.uniform(5, 11)
            charging_time = random.uniform(6, 9)
        this_car_name = "Car "+str(i+1)
        this_car =  Electric_Car(this_car_name,
            charging_time)
        activate(this_car,this_car.visit_station( ),at=
            arrival_time)

    simulate(until=simulation_time)

    average_waiting_times.append(car_monitor.mean())
    if monte_carlo == False:
        print("\nCars Entered = %3.0f" % car_monitor.
            count())
        print("Cars Served = %3.0f" % cars_served)
        print ("Mean Waiting Time = %3.1f minutes" %
            car_monitor.mean())
        print ("Waiting Time    = %3.1f minutes" %
            car_monitor.var()**.5)
        print ('Utilization = %3.1f ' % charging_station.
            actMon.timeAverage())
        low = car_monitor.mean() - 1.96*car_monitor.var()
            **.5/(cars_served**.5)
        high = car_monitor.mean() + 1.96*car_monitor.var
            ()**.5/(cars_served**.5)
        print ('95%% Confidence Interval = %3.1f - %3.1f'
             % (low, high))
        print ('Average Queue Length = %4.1f ' %
            charging_station.waitMon.timeAverage())
        output_file.write("%2.0f %6.2f %6.2f" % (
            iteration, car_monitor.mean(),
            charging_station.actMon.timeAverage()) + '\n')

    if monte_carlo == True: output_file.write("%2.0f %6.3
        f %6.3f" % (iteration, car_monitor.mean(),
        charging_station.actMon.timeAverage()) + '\n')

output_file.close()

if monte_carlo == True:
    print ('M-C mean waiting time = %4.3f ' %
        statistics.mean(average_waiting_times))
```

```
    print ('M-C variance of waiting time = %4.3f ' %
       statistics.variance(average_waiting_times))
```

B.2 CONTINUOUS SYSTEM MODELS

B.2.1 INTEGRATION USING EULER'S METHOD

This example demonstrates Euler's method of integration. It integrates velocity (dx/dt) at each dt in a time loop to determine position (x).

```
# demonstrate Euler's method of integration to
   calculate position over time

time = 0
dt = .25 # Timestep in minutes
position = 5 # Initial position (kilometers)
velocity = .8 # Velocity (kilometers/minute)

print ("Initial position: %6.2f, velocity: %6.2f" % (
   position, velocity))

# header for time output
print ("Time  Position  Velocity")

# run for 10 minutes
while time < 10:
    time = time + dt
    # integrate velocity to get new position: x(t) = x(
       t-1) + dx/dt * dt
    position = position + velocity*dt
    print ("%6.2f %6.2f %6.2f" % (time, position,
       velocity))

print ("Final position: %6.2f" % position)
```

B.2.2 INTEGRATION USING THE RUNGE-KUTTA METHOD

This example implements the more sophisticated fourth-order Runge-Kutta method of integration. The function *Runge˙Kutta* is called in a time loop to update the position and velocity as a state vector of a damped spring.

```
# calculate spring dynamics using Runge-Kutta method of
    integration
```

```
def Runge_Kutta(x, v, a, dt):
    """Runge-Kutta integration function returns final (
       position, velocity) tuple after
    time dt has passed.  Implements fourth order Runge-
       Kutta method.

    x: initial position
    v: initial velocity
    a: acceleration function a(x,v,dt)
    dt: timestep"""
    x1 = x
    v1 = v
    a1 = a(x1, v1, 0)

    x2 = x + 0.5*v1*dt
    v2 = v + 0.5*a1*dt
    a2 = a(x2, v2, dt/2.0)

    x3 = x + 0.5*v2*dt
    v3 = v + 0.5*a2*dt
    a3 = a(x3, v3, dt/2.0)

    x4 = x + v3*dt
    v4 = v + a3*dt
    a4 = a(x4, v4, dt)

    xf = x + (dt/6.0)*(v1 + 2*v2 + 2*v3 + v4)
    vf = v + (dt/6.0)*(a1 + 2*a2 + 2*a3 + a4)
    return xf, vf

def acceleration(x, v, dt):
    """Determines acceleration from current position,
    velocity, and timestep. This acceleration
    function models a spring."""
    stiffness = 1
    damping = -0.005
    return -stiffness*x - damping*v

t = 0
dt = 1.25 # Timestep in seconds
state = 50, 5 # Position, velocity

print ("Initial position: %6.2f, Velocity: %6.2f" %
   state)
```

```
# label for time loop output
print ("  time  position velocity")

# Run for 10 seconds
while t < 10:
    t += dt
    state = Runge_Kutta(state[0], state[1],
       acceleration, dt)
    print ("%6.2f %6.2f %6.2f" % (t, state[0], state
       [1]))

print ("Final position: %6.2f, velocity: %6.2f" % state
   )
```

Glossary

Activity: A task or processing delay that requires resources for a duration. The start and end times of activities are events.

Attribute: A property or characteristic of one or more entities.

Constant: A quantity or data item whose value does not change during a simulation run.

Continuous Model: A mathematical model whose output variables change in a continuous manner with respect to time. Contrast with: Discrete Event Model.

Continuous System: A system for which the state variables change continuously with respect to time.

Deterministic: Characteristic of a process, model, simulation, or variable whose outcome, result, or value is certain.

Deterministic Model: A model in which the results are determined through known relationships among the states and events and in which a given input will always produce the same output. Contrast with: Stochastic Model.

Discrete Event Model: A mathematical model whose output variables change from one value to another at discrete event times. Contrast with: Continuous Model.

Discrete System: A system for which the state variables change instantaneously at separate points in time.

Entity: An object in a system that has attributes whose motion may result in an event.

Environment: 1) The external surroundings or conditions that a modeled system operates within, 2) a simulation tool execution construct for dynamic modeling.

Event: An occurrence causing changes in the state of a discrete system.

Initial Condition: The values assumed by the variables in a system, model, or simulation at the beginning of some specified time period.

Measure of Effectiveness (MOE): A qualitative or quantitative measure of a characteristic indicating the degree to which a desired effect is obtained or an operational objective or requirement is met under specified conditions.

Model: A physical, mathematical, or otherwise logical representation of a system, entity, phenomenon, or process.

Monte Carlo: A simulation in which random statistical sampling techniques are employed.

Probabilistic Model: See Stochastic Model.

Process: The step-by-step actions or behavior of entities that flow through the system.

Random: Pertaining to a process or variable whose outcome or value depends on chance.

Resource: A commodity used by entities while being served in a system.

Queue: A set of entities waiting to be serviced.

State: A static view of a system described by the values of its variables.

Static Model: A model in which no change occurs with respect to time.

Steady State: A situation in which a model, process, or device exhibits stable behavior independent of time.

Stochastic Model: A model in which the results are determined by using one or more random variables to represent uncertainty. Contrast with: Deterministic Model.

Simulation: A method for executing a model over time.

System: A set of interacting parts that form a connected whole accomplishing a specific function or set of functions.

Validation: Determining the degree to which a model and simulation is an accurate representation of a real system or process within its intended purpose and scope.

Variable: A quantity or data item whose value can change.

Verification: Determining that a model or simulation implementation accurately represents the developer's conceptual description and specification.

See [23] for a more extensive glossary of modeling and simulation terms.

Bibliography

[1] Argonne National Laboratory. The Repast Suite: https://repast.github.io, 2015.

[2] J. Banks and J. Carson. *Discrete-Event System Simulation*. Prentice Hall, 1996.

[3] E. Z. Berglund. Using agent-based modeling for water resources planning and management. *Journal of Water Resources Planning and Management*, 141(11), 2015.

[4] K. G. Cooper and T. W. Mullen. Swords and plowshares: The rework cycle of defense and commercial software development projects. *American Programmer*, 6(5), 1993.

[5] J. W. Forrester. *Industrial Dynamics*. Cambridge, MA: MIT Press, 1961.

[6] J. W. Forrester. *Principles of Systems*. Cambridge, MA: MIT Press, 1968.

[7] S. Fortmann-Roe. Insight Maker: A general-purpose tool for web-based modeling and simulation. *Simulation Modelling Practice and Theory*, 47:28–45, 2014.

[8] Give Team,. Insight Maker: http://insightmaker.com/, 2016.

[9] C.R. Hicks and K. V. Turner. *Fundamental Concepts in the Design of Experiments, 5th Edition*. Oxford University Press, 1999.

[10] D. Houston. A generalized duration forecasting model of test-and-fix cycles. *Journal of Software: Evolution and Process*, 26(10):877–889, 2014.

[11] C. C. Hsu and B. A. Sandford. The delphi technique: Making sense of consensus. practical assessment. *Research & Evaluation*, 12(10), 2007.

[12] D. W. Hubbard. *How to Measure Anything,3rd ed.* Hoboken, NJ: John Wiley and Sons, 2014.

[13] Imagine That Inc. *ExtendSim User Guide*. Imagine That Inc., 2013.

[14] INCOSE. *INCOSE Systems Engineering Handbook, Version 4.0*. John Wiley and Sons, Hoboken, NJ, 2015.

[15] JaamSim Development Team. Jaamsim: Discrete-event simulation software: http://jaamsim.com, 2016. Version 2016-14.

[16] B. Khoshnevis. *Discrete Systems Simulation*. McGraw-Hill, New York, NY, 1994.

[17] A. M. Law. *Simulation Modeling and Analysis*. McGraw-Hill Higher Education, 5th edition, 2014.

[18] R. J. Madachy. *Software Process Dynamics*. Hoboken, NJ: John Wiley and Sons, 2008.

[19] MathWave. Statassist: http://www.mathwave.com/help/easyfit/html/tools/assist.html/, 2017.

[20] N Matloff. Introduction to Discrete-Event Simulation and the SimPy Language: http://heather.cs.ucdavis.edu/~matloff/156/%20PLN/DESimIntro.pdf, 2008.

[21] D. C. Montgomery. *Design and Analysis of Experiments, 8th Edition*. New

York, NY: John Wiley and Sons, 2012.

[22] M. J. North, N. T. Collier, and J. R. Vos. Experiences creating three implementations of the repast agent modeling toolkit. *ACM Trans. Model. Comput. Simul.*, 16(1):1–25, January 2006.

[23] United States Department of Defense. DoD Modeling and Simulation (M&S) Glossary. (DoD 5000.59-M), 1998.

[24] G. P. Richardson. System dynamics: Simulation for policy analysis from a feedback perspective. In Fishwich and Luker, editors, *Qualitative Simulation Modeling and Analysis*. Springer-Verlag, 1991.

[25] G. P. Richardson and A. L. Pugh. *Introduction to System Dynamics Modeling with DYNAMO*. Productivity Press, Cambridge, MA, 1981.

[26] K. L. Sanford Bernhardt and S. McNeil. Agent-based modeling: Approach for improving infrastructure management. *Journal of Infrastructure Systems*, 14(3), 2008.

[27] SimPy. SimPy Simulation Package: `http://simpy.sourceforge.net`, 2016.

[28] J. Son, E. M Rojas, and S. Shin. Application of agent-based modeling and simulation to understanding complex management problems in cem research. *Journal of Civil Engineering and Management*, 21(8):998–1013, 2015.

[29] J. Sterman. *Business Dynamics: Systems Thinking and Modeling for a Complex World*. Boston MA: Irwin McGraw-Hill, 2000.

Index

About the Authors

Dr. Raymond Madachy is an Associate Professor in the Systems Engineering Department at the Naval Postgraduate School. He has 30 years of experience working in industry, academia, and consulting in technical and management positions. His research interests include modeling and simulation of processes for architecting and engineering of complex systems; system total ownership cost modeling and tradespace analysis; systems and software measurement; integrating systems engineering and software engineering disciplines; and integrating empirical research with process simulation. He has over 100 publications with three previous books including *Software Process Dynamics*.

Dr. Daniel Houston is a Sr. Project Leader at The Aerospace Corporation in Los Angeles, California, U.S.A. His work is applying quantitative methods, particularly using statistics and simulation, to software engineering. His industrial background includes software development, Six Sigma Black Belt, and software measurement. He received M.S. and Ph.D. degrees in Industrial Engineering at Arizona State University. His publications include work on statistical modeling and simulation of software development processes, software process improvement, and the management of software projects, particularly the aspects of risk, product quality, and economics.

#0002 - 050817 - C170 - 234/156/9 - PB - 9781498753098